THE TEN FACES
OF INNOVATION

TOM KELLEY is the general manager of IDEO – the design firm that has developed thousands of innovative products and services, ranging from Kraft's supply chain efficiencies to the Palm V personal organizer.

JONATHAN LITTMAN is a contributing editor at *Playboy* and author of several books on business, crime and sport.

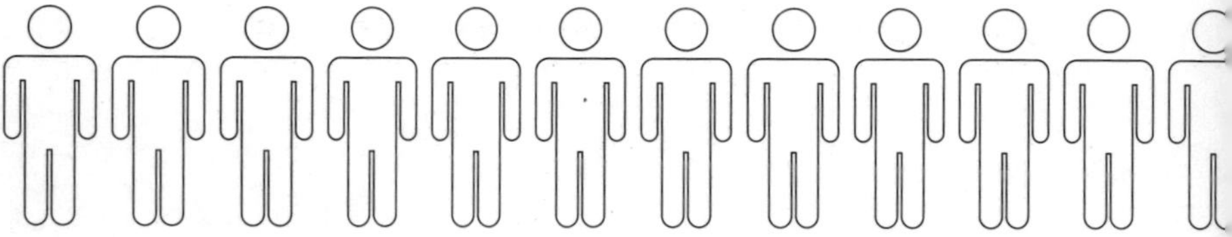

THE TEN FACES OF INNOVATION

IDEO'S STRATEGIES FOR BEATING THE DEVIL'S ADVOCATE &
DRIVING CREATIVITY THROUGHOUT YOUR ORGANIZATION

TOM KELLEY

WITH JONATHAN LITTMAN

PROFILE BOOKS

First published in Great Britain in 2006 by
PROFILE BOOKS LTD
3A Exmouth House
Pine Street
Exmouth Market
London EC1R 0JH
www.profilebooks.com

First published in the United States in 2006 by
CURRENCY BOOKS
A division of Random House, Inc.

1 3 5 7 9 10 8 6 4 2

Printed and bound in Great Britain by
Clays, Bungay, Suffolk

A CIP catalogue record for this book is available from the British Library.

ISBN-10: 1 86197 806 5
ISBN-13: 978 1 86197 806 6

To my family
—the Okusan, the Ballerina, and the Bonzuppe—
for their patience, understanding, and love.

CONTENTS

ACKNOWLEDGMENTS

If you prefer to cling to the romantic notion of a lonely author toiling away like a starving artist in a dimly lit garret, you might want to skip this page. Although getting the words down on paper can still be a very lonely task, of course, making a book actually takes quite a crowd. More than a hundred people worked on the book you now hold in your hands, though I will not attempt to mention them all one-by-one.

So here are a dozen individuals or teams I'd like to single out for special thanks:

Scott Underwood applied his encyclopedic knowledge of words in giving me advice on syntax, grammar, and elements of style. I have learned more about the nuances of the English language from Scott than I have from any professor. I owe him not only my friendship but also a debt of gratitude.

Olympic athlete and journalist Brigit Finn took time away from her busy schedule at *Business 2.0* to investigate dozens of innovation stories, and I occasionally wonder if it was her influence that gave the book so many metaphors from the Olympic Games.

Brendan Boyle and David Haygood distinguished themselves as the most prolific contributors from among the dozens of in-person interviews and thousands of e-mail responses on innovation topics. If they appear more often than others, it is because they provided a constant stream of useful input—even after I stopped asking for it.

Marc Hershon generously lent me his "think tank and tiki lounge" in Sausalito as a writer's hideout so I could periodically escape the continuous-interrupt environment of my desk at IDEO. Marc's space actually has a *door*—though I never had occasion to close it.

Hunter Lewis Wimmer used a blend of design talents and diplomacy skills to turn my vague requests into tangible design elements like the cover and chapter intros. Hunter survived the experience and left his imprimatur on the physical appearance of the book.

Lynn Winter searched out, created, or acquired nearly every photographic image you'll see between these covers, lending her energy and perseverance to the project at a time when I felt nearly exhausted by it.

Tim Brown, David Strong, and Peter Coughlan managed to remain patient and supportive with a book project that took *way* more time away from my day-to-day responsibilities than I had originally imagined. My vision of doing the book nights and weekends was perfectly accurate but incomplete, since the work spilled over into a couple of hundred weekdays, too.

In the literary world, I had lots of help (again) from my agent and mentor Richard Abate, as well as Doubleday executive editor Roger Scholl. Also, Chris Fortunato and his magical book-packaging team raced through the production cycle—while jumping hurdles—about twice as fast as I expected.

My brother, David, ultimately made this book possible, not only by starting IDEO and seeding many of the ideas contained here, but also by being my greatest source of advice and guidance for half a century and counting. In addition to everything else, he lent me his Stanford office—which became another secret getaway spot for writing and editing. I know I will never be able to repay him, and—even better—that he doesn't expect me to.

Jon Littman's contribution was so great that I am not sure whether it's appropriate to thank him here, since he could rightly have his own page of acknowledgments. In many parts of the written manuscript, it's impossible for me to separate my efforts from Jon's, and we've had an intensely close collaboration that has now spanned over six years. He faithfully stuck with the project, even while others were luring him away with racy alternatives.

My wife, Yumiko, didn't work on the project directly, but she did take on plenty of extra parenting roles for the past eighteen months so I could go on this book adventure. Yumi and the family made a lot of sacrifices during that time, and still supported me all the way. I try never to take them for granted.

As for everyone else, you know who you are: trusted advisors like Whitney Mortimer and Debbe Stern; tireless supporters like Joani Ichiki and Kathleen Hughes; spontaneous helpers like Katie Clark and Marguerite Rigoglioso; Transformation colleagues like Ilya Prokopoff, Charles Warren, and Hilary Hoeber; thoughtful reviewers like Paul Bennett, Roby Stancel, Diego Rodriguez, and Jed Morley; and sources of inspiration like Tom Peters, Bob Sutton, Malcolm Gladwell, Ron Avitzur, Stefan Thomke, Stephen Denning, Seth Godin, and the Group of 33.

Thanks, everybody. I hope you're happy with the finished book.

INTRODUCTION
Beyond the Devil's Advocate

We've all been there. The pivotal meeting where you push forward a new idea or proposal you're passionate about. A fast-paced discussion leads to an upwelling of support that seems about to reach critical mass. And then, in one disastrous moment, your hopes are dashed when someone weighs in with those fateful words: *"Let me just play Devil's Advocate for a minute . . ."*

Having invoked the awesome protective power of that seemingly innocuous phrase, the speaker now feels entirely free to take potshots at your idea, and does so with complete impunity. Because they're *not really* your harshest critic. They are essentially saying, "The Devil made me do it." They're removing themselves from the equation and side-stepping individual responsibility for the verbal attack. But before they're done, they've torched your fledgling concept.

The Devil's Advocate gambit is extraordinary but certainly not uncommon, since it strikes so regularly in the project rooms and board-rooms of corporate America. What's truly astonishing is how much punch is packed into that simple phrase. In fact, the Devil's Advocate may be the biggest innovation killer in America today. What makes this negative persona so dangerous is that it is such a subtle threat. Every day, thousands of great new ideas, concepts, and plans are nipped in the bud by Devil's Advocates.

Why is this persona so damning? Because the Devil's Advocate encourages idea-wreckers to assume the most negative possible perspective, one that sees only the downside, the problems, the disasters-in-waiting. Once those floodgates open, they can drown a new initiative in negativity.

Why should you care? And why do I believe this problem is so important? Because innovation is the lifeblood of all organizations, and the Devil's Advocate is toxic to your cause. This is no trivial matter. There is no longer any serious debate about the primacy of innovation to the health and future strength of a corporation. Even the staid British publication *The Economist* recently claimed, "Innovation is now recognized as the single most important ingredient in any modern economy."

And what *The Economist* said about nations is equally true about organizations. In the four years since *The Art of Innovation*, my first book about our practices at IDEO, I have worked with clients from Singapore to San Francisco to São Paulo. At the same time, the scope of our work has expanded to include industries as far-flung as health care services, retailing, transportation, financial services, consumer packaged goods, and food and beverage. I have witnessed firsthand how innovation has become recognized as a pivotal management tool across virtually all industries and market segments. And while we at IDEO used to spend the majority of our time in the world of product-based innovation, we have more recently come around to seeing innovation as a tool for transforming the entire culture of organizations. Sure, a great product can be one important element in the formula for business success, but companies that want to succeed in today's competitive environment need much more. They need innovation at every point of the compass, in all aspects of the business and among every team member. Building an environment fully engaged in positive change, and a culture rich in creativity and renewal, means creating a company with 360 degrees of innovation. And companies that want to succeed at innovation will need new insights. New viewpoints. And new roles.

> Even the staid British publication *The Economist* recently claimed, "Innovation is now recognized as the single most important ingredient in any modern economy."

There is growing recognition that fostering a culture of innovation is critical to success, as important as mapping out competitive strategies or maintaining good margins. A recent Boston Consulting Group survey covering nearly fifty countries and all sorts of businesses reported that nine out of ten senior executives believe generating growth through innovation is essential for success in their industry. Where business magazines once ranked companies primarily by sales, growth, and profit, publications are now ranking corporations on their innovation track record. And while acquisitions can yield synergy, and reengineering can streamline operations, a culture of innovation may be the ultimate fuel for long-term growth and brand development. Having optimized operations and finances, many companies are now recognizing that growth through innovation is their best strategy to compete in a world marketplace in which some of the players may have lower-cost resources. As my friend Tom Peters would say, you can't shrink your way to greatness. One way to look at the current pressure-cooker of international business is as a fierce competition, where you win through innovation or lose the game. Today, companies are valued less for their current offerings than for their ability to change and adapt and dream up something new. Whether you sell consumer electronics or financial services, the frequency with which you must innovate and replenish your offerings is rapidly increasing.

Serial Innovation Success

As I was completing this book, Google, already the world's leading search engine, was innovating at a breakneck pace, rolling out a new service capability or acquisition practically every month—everything from searching rare books in the world's greatest libraries to viewing aerial photographs of any location or skimming through transcripts of last night's TV shows. Until Google introduced Desktop Search, I had thought of it only as a firm to help me search the Web. Now they've convinced me I'll soon be using a search engine to wade through all my own data as well.

Google, of course, is not alone in such rapid-fire innovation. Plenty of companies in widely divergent industries have distinguished themselves as serial innovators. Here are a few favorites that come to mind:

○ W. L. Gore & Associates, most famous for its breathable Gore-Tex fabrics, not only manufactures a tremendous breadth of products—everything from guitar strings to artificial blood vessels—it also distinguishes itself through its egalitarian, team-based organization. Eschewing bosses and job descriptions, Gore creates idea-friendly environments that work to generate a continual stream of clever innovations. Gore was recently cited as "the most innovative company in America," and is ranked among the best places to work in Germany, Italy, the United States, and the United Kingdom.

○ The Gillette Company grabbed enormous market share over the years with a series of newer-and-better shaving systems like the Sensor and Mach III razors. Far from resting on its laurels, the firm recently poured its considerable resources into an even more ambitious project, the motorized M3 Power razor. Along the way, Gillette has developed a culture of continuous innovation to stay a step ahead of its competitors.

○ The unique German retailer Tchibo started in the 1950s as a simple coffee shop, but has transcended its roots to become an international merchandising sensation. Tchibo is like Starbucks meets Brookstone, combining a stand-up café with an eclectic collection of ever-changing products. Part of its success formula is "A new experience every week," with a completely new line of inventory (everything from bicycles to lingerie) arriving and selling in huge quantities for just seven days. Tchibo reports, for example, that the week their stores featured telescopes they sold more telescopes in seven days than had been sold the previous *year* in all of Germany. The company continues to introduce a completely new merchandising theme fifty-two times a year and generates impressive sales throughout Europe.

The Human Touch

The Ten Faces of Innovation is a book about innovation with a human face. It's about the individuals and teams that fuel innovation inside great organizations. Because all great movements are ultimately human-powered. Archimedes said, "Give me a place to stand and a lever long enough and I can move the world." The innovation personas described in the next ten chapters are not necessarily the most powerful people you will ever meet. They don't have to be. Because each persona brings its own lever, its own tools, its own skills, its own point of view. And when someone combines energy and intelligence with the right lever, they can generate a remarkably powerful force. Make sure they have a place on your team. Together you can do extraordinary things.

At IDEO, we believe that innovators focus on the verbs. They're proactive. They're energetic. Innovators set out to create, to experiment, to inspire, to build on new ideas. Our techniques may at times seem unusual, but the results can be truly extraordinary.

All good working definitions of innovation pair ideas with action, the spark with the fire. Innovators don't just have their heads in the clouds. They also have their feet on the ground. 3M, one of the first companies to fully embrace innovation as the essence of its corporate brand, defines it as "New ideas—plus action or implementation—which result in an improvement, a gain, or a profit." It is not enough to just have a good idea. Only when you *act*, when you *implement*, do you truly innovate. Ideas. Action. Implementation. Gain. Profit. All good words, of course, but there's still one piece left out. *People*. That's why I prefer the Innovation Network's definition: "*People* creating value through the implementation of new ideas." The classic 3M definition might leave you with the impression that, as a bumper sticker might put it, "Innovation Happens." But unfortunately, there's no spontaneous combustion in the business world. Innovation is definitely not self-starting or self-perpetuating. People make it happen through their imagination, willpower, and perseverance. And whether you are a team member, a group leader, or an executive, your only real path to innovation is through people. You can't really do it alone.

This is a book about people. More specifically, it is about the roles people can play, the hats they can put on, the personas they can adopt. It is not about the luminaries of innovation like Thomas Edison, or even celebrity CEOs like Steve Jobs and Jeffrey Immelt. It is about the unsung heroes who work on the front lines of entrepreneurship in action, the countless people and teams who make innovation happen day in and day out.

The ten core chapters of this book highlight ten people-centric tools developed at IDEO that you might call talents or roles or personas for innovation. Although the list does not presume to be comprehensive, it does aspire to expand your repertory. We've found that adopting one or more of these roles can help teams express a different point of view and create a broader range of innovative solutions.

By developing some of these innovation personas, you'll have a chance to put the Devil's Advocate in his place. So when someone says, "Let me play Devil's Advocate for a minute" and starts to smother a fragile new idea with negativity, someone else in the room may be emboldened to speak up and say, "Let me be an Anthropologist for a moment, because I personally have watched our customers suffering silently with this issue for months, and this new idea just might help them." And if that one voice gives courage to others, maybe someone else will add, "Let's think like an Experimenter for a moment. We could prototype this idea in a week and get a sense of whether we're onto something good." Or someone else could volunteer to be a Hurdler, and pledge to get the team some seed funding for an exploration of the concept. The Devil's Advocate may never go away, but on a good day, the ten personas can keep him in his place. Or tell him to go to hell.

One important caveat. My feelings about Devil's Advocates should not be interpreted as some sort of endorsement for a "yes-man culture." IDEO has always believed in constructive criticism and free debate. Actually, strong innovation roles can lead to more critical thinking, as team members develop a broader perspective

> The Devil's Advocate may never go away, but on a good day, the ten personas can keep him in his place. Or tell him to go to hell.

from which to view projects. But the Devil's Advocate seldom takes a real stand, preferring to tear down an idea with clever criticism, and often exhibiting the mean-spirited negativity associated with that role. Meanwhile, the innovation roles are intended to encourage people to stand up for what they believe in.

So who are these personas? Many already exist inside of large companies, though they're often underdeveloped or unrecognized. They represent latent organizational ability, a reservoir of energy waiting to be tapped. We all know plenty of bright, capable people who would like to make a bigger contribution, team members whose contributions don't quite fit into traditional categories like "engineer" or "marketer" or "project manager."

In a postdisciplinary world where the old descriptors can be constraining, these new roles can empower a new generation of innovators. They give individuals permission to make their own unique contribution to the social ecology and performance of the team. Here's a brief introduction of the personas:

The Learning Personas

Individuals and organizations need to constantly gather new sources of information in order to expand their knowledge and grow, so the first three personas are *learning roles*. These personas are driven by the idea that no matter how successful a company currently is, no one can afford to be complacent. The world is changing at an accelerated pace, and today's great idea may be tomorrow's anachronism. The learning roles help keep your team from becoming too internally focused and remind the organization not to be so smug about what you "know." People who adopt the learning roles are humble enough to question their own worldview, and in doing so they remain open to new insights every day.

1 The Anthropologist brings new learning and insights into the organization by observing human behavior and developing a deep understanding of how people interact physically and emo-

tionally with products, services, and spaces. When an IDEO human-factors person camps out in a hospital room for forty-eight hours with an elderly patient undergoing surgery—as described in Chapter 1—she is living the life of the Anthropologist and helping to develop new health care services.

2 The Experimenter prototypes new ideas continuously, learning by a process of enlightened trial and error. The Experimenter takes calculated risks to achieve success through a state of "experimentation as implementation." When BMW bypassed all its traditional advertising channels and created theater-quality short films for bmwfilms.com, no one knew whether the experiment would succeed. Their runaway success, which underscores the rewards that flow to Experimenters, is detailed in Chapter 2.

3 The Cross-Pollinator explores other industries and cultures, then translates those findings and revelations to fit the unique needs of your enterprise. When an open-minded Japanese businesswoman travels 5,000 miles to find inspiration for a new brand, she finds a concept an ocean away that sparks a billion-dollar retail empire, and demonstrates the leverage of a Cross-Pollinator. You'll hear her story in Chapter 3.

The Organizing Personas

The next three personas are *organizing roles*, played by individuals who are savvy about the often counterintuitive process of how organizations move ideas forward. At IDEO, we used to believe that the ideas should speak for themselves. Now we understand what the Hurdler, the Collaborator, and the Director have known all along: that even the best ideas must continuously compete for time, attention, and resources. Those who adopt these organizing roles don't dismiss the process of budget and resource allocation as "politics" or "red tape." They recognize it as a complex game of chess, and they play to win.

4 **The Hurdler** knows the path to innovation is strewn with obstacles and develops a knack for overcoming or outsmarting those roadblocks. When the 3M worker who invented Scotch tape decades ago had his idea initially rejected, he refused to give up. Staying within his $100 authorization limit, he signed a series of $99 purchase orders to pay for critical equipment needed to produce the first batch. His perseverance paid off, and 3M has reaped billions of dollars in cumulative profits because an energetic Hurdler was willing to bend the rules.

5 **The Collaborator** helps bring eclectic groups together, and often leads from the middle of the pack to create new combinations and multidisciplinary solutions. When a customer-service manager wins over a skeptical corporate buyer to the idea of brainstorming new forms of cooperation, and the resulting new program doubles their sales, he's playing the role of a very successful Collaborator.

6 **The Director** not only gathers together a talented cast and crew but also helps to spark their creative talents. When a creative Mattel executive assembles an ad hoc team and dubs them "Platypus," launching a novel process that creates a $100 million toy platform in three months, she is a role model for Directors everywhere. Her story is told in Chapter 6.

The Building Personas

The four remaining personas are *building roles* that apply insights from the learning roles and channel the empowerment from the organizing roles to make innovation happen. When people adopt the building personas, they stamp their mark on your organization. People in these roles are highly visible, so you'll often find them right at the heart of the action.

7 The Experience Architect designs compelling experiences that go beyond mere functionality to connect at a deeper level with customers' latent or expressed needs. When an ice cream shop turns the preparation of a frozen dessert into a fun, dramatic performance, it is designing a successful new customer experience. The premium prices and marketing buzz that follow are rewards associated with playing the role of the Experience Architect.

8 The Set Designer creates a stage on which innovation team members can do their best work, transforming physical environments into powerful tools to influence behavior and attitude. Companies like Pixar and Industrial Light & Magic recognize that the right office environments can help nourish and sustain a creative culture. When a business team doubles its usable output after reinventing its space and a sports team discovers a renewed winning ability in a brand-new stadium, they are demonstrating the value of the Set Designer. Organizations that tap into the power of the Set Designer sometimes discover remarkable performance improvements that make all the space changes worthwhile.

9 The Caregiver builds on the metaphor of a health care professional to deliver customer care in a manner that goes beyond mere service. Good Caregivers anticipate customer needs and are ready to look after them. When you see a service that's really in demand, there's usually a Caregiver at the heart of it. A Manhattan wine shop that teaches its customers how to enjoy the pleasures of wine without ever talking down to them is demonstrating the Caregiver role—while earning a solid profit at the same time.

10 The Storyteller builds both internal morale and external awareness through compelling narratives that communicate a fundamental human value or reinforce a specific cultural trait. Companies from Dell to Starbucks have lots of corporate legends that support their brands and build camaraderie within their teams. Medtronic, celebrated for its product innovation and

consistently high growth, reinforces its culture with straight-from-the-heart storytelling from patients' firsthand narratives of how the products changed—or even saved—their lives.

The appeal of the personas is that they work. Not in theory or in the classroom but in the unforgiving marketplace. IDEO has battle-tested them thousands of times in a real-world laboratory for innovation. Every year, we work on literally hundreds of innovation projects. And where once the bulk of our clients were start-ups or technology companies, today some of our biggest clients are progressive leaders of the Fortune 100. They seek us out not just for help with a single innovation but for a series of innovations. They come to us to tap into the insights and energy of a talented team, adept at playing roles like Cross-Pollinator, Anthropologist, and Experimenter.

Transforming Innovation

The Ten Faces of Innovation is designed to help you bring the human elements of innovation to the workings of your enterprise. In giving innovation a face, I've also tried to give it a personality. And I've had a lot of help, thanks not only to my brother, David, who founded the firm, but also to the hundreds of talented IDEO designers, engineers, and human factors people who have paved the way over the last twenty-seven years. It's my hope that this book pays them tribute by shining a light on the essential approaches, personas, and roles that nourish innovation.

The Ten Faces of Innovation is about how people and teams put into practice methods and techniques that infuse an enterprise with a continuous spirit of creative evolution. Successful businesses build fresh innovation strategies into the fabric of their operations. They do it year-round and in widely differing parts of their enterprises. When the team's creative engine is running at top speed, the momentum and synergy can keep a company ahead through bad times and good.

In an increasingly competitive global marketplace, this book is about seizing the innovation opportunities in a company, an industry, a region, even a nation. It's about developing the personas of your team to maximize its influence. The right innovation project at the right time can spur a companywide movement, generating an afterglow that permeates the workplace—sparking a culture of innovation that takes on a life of its own.

The proof, as they say, is in the pudding. In the following chapters, you'll find ample evidence of the transforming power of a culture of innovation. You'll find companies where innovation is no longer merely about generating compelling new products and services. Companies where the creative process itself—how they work, inspire, and collaborate—has developed a remarkable energy that keeps the organization moving forward.

As you get to know the ten personas, keep in mind that they're not inherent personality traits or "types" that are permanently attached to one (and only one) individual on the team. A persona is not about your predetermined "business DNA." These innovation roles are available to nearly anyone on your team, and people can switch roles, reflecting their multifaceted capabilities.

This nimble contextual switching from role to role may sound a bit complicated, but you are already probably very good at it. For example, I play at least half a dozen roles every day, including husband, father, brother, IDEOer, author, speaker, mentor, and Transformation team member. Completely immersed in one of my business roles, I get an urgent phone call from my son, and I switch instantly into father role. In doing so, I change my attitude, my tone of voice, my patience level, and even my thought patterns. Staying in one role when I need to be in the other would be inappropriate and ineffective. Worse yet, it could damage relationships or even my career. But getting the role just right can be very rewarding.

It's the same with innovation roles. We have too many people out there playing Devil's Advocate when they should be in a learning role like the Anthropologist, when they should be invoking an organizing role like the Collaborator, when they should be adopting a building role like the Experience Architect. The innovation roles give you a chance to broaden your creative range, with the flexibility to pick the right role for the right challenge. The innovation roles offer a new vocabulary, sparking a fresh discourse that invites team members to make their own unique contributions to the success of the enterprise.

And like a Method actor immersing himself in a new role, you may find that walking in the shoes of a new persona changes your attitude and outlook, even your behavior. If it opens you up to new thought patterns, the new role may help you achieve personal and professional growth. And thinking of the ten innovation elements as personas rather than tools reminds us that innovation is a full-time endeavor for all modern organizations, not just a task to be checked off periodically. The personas are about "being innovation" rather than merely "doing innovation." Taking on one or more of these roles is a conscious step toward becoming an innovator in your daily life.

When you begin building your team, remember there is no set formula for using the personas. People can take on multiple roles. You need not have a one-to-one mapping of teams to personas, and you certainly don't need ten people on every team. It's unlikely every team will have every persona represented. Conversely, this isn't Hollywood, and no one wants to be typecast. You might find yourself wearing the hats of two or three personas as you move from one project to the next.

Some of these roles will undoubtedly fit you better than others. You may be a born Cross-Pollinator or a nimble Experimenter. You may also find you're a better Anthropologist than you thought possible. This isn't a competition between the individual innovation roles. It's a team effort to expand the overall potential of your organization. Increasing your skills in just two or three roles can make a critical difference. *The Ten Faces of Innovation* is about inviting you to broaden your color palette. Maybe you've always favored blue and green, but if you open these pages and try a few brushstrokes of purple, you might be amazed at the results. So take up your brush and let it fly.

The canvas is waiting.

CHAPTER 1
The Anthropologist

> The real act of discovery consists not in finding new
> lands, but in seeing with new eyes. —MARCEL PROUST

I **f I could choose just one persona,** it would be the Anthropologist. I have the fervor of a convert on this one, because back when I
joined the tiny firm that became IDEO, there were no Anthropologist
roles. Experimenters, yes. Even a few Cross-Pollinators. But no one had
yet adopted the persona that has since become the cornerstone of our
work. When the notion of applying anthropology first came to IDEO
in 1991, I wish I could say I was a visionary, instantly recognizing it
as the future of the firm. In fact, the opposite was true. I remember
saying to my brother David at the time, "Now here's a sweet job. All
these bright people with Ph.D.s have to do is *watch* people. They take
a few pictures while they're at it, maybe a video clip or two, and then
come back and tell us about what they saw. That hardly sounds like
work at all." Meanwhile, our engineers were hunkered down at their
CAD machines, trying to create electronic products that could survive
a four-foot drop onto concrete without breaking. Now *that* seemed like
hard work.

But in the intervening years, I have come around 180 degrees on
the role of the Anthropologist in our firm. Far from being some fluffy,
esoteric process of questionable value, the Anthropologist role is the
single biggest source of innovation at IDEO. Like most of our client
companies, we have lots of great problem-solvers. But you have to know
what problem to solve. And people filling the Anthropologist role can
be extremely good at reframing a problem in a new way—informed by

their insights from the field—so that the right solution can spark a breakthrough.

So what makes Anthropologists so valuable? At IDEO, people in this role typically start with a very solid grounding in the social sciences, coming to us with advanced degrees in subjects like cognitive psychology, linguistics, or anthropology. But what's apparent when you work with them is not their academic knowledge so much as a sense of informed intuition, akin to what Harvard Business School professor Dorothy Leonard calls "Deep Smarts." Although no IDEO Anthropologist has ever given me a unified theory of their role, I have noticed half a dozen distinguishing characteristics. Some are strategic and some are quite tactical:

Anthropologists seek out epiphanies through a sense of "Vuja De."

1 **Anthropologists practice the Zen principle of "beginner's mind."** Even with extensive educational backgrounds and lots of experience in the field, people in the Anthropologist role seem unusually willing to set aside what they "know," looking past tradition and even their own preconceived notions. They have the wisdom to observe with a truly open mind.

2 **Anthropologists embrace human behavior with all its surprises.** They don't judge, they observe. They empathize. Lifelong students of human behavior develop a genuine love of watching and talking to people that cannot be faked. The skills and techniques of cultural anthropology can be learned by anyone, but the people drawn to this role usually find it intrinsically rewarding, which is another way of saying that they love their work.

3 **Anthropologists draw inferences by listening to their intuition.** Both the business-school curriculum at prominent universities and on-the-job learning in the corporate world focus on exercising our left-brain analytical skills. They sharpen our deduc-

tive reasoning powers, what Guy Claxton calls the "d-brain" in his intriguing book *Hare Brain, Tortoise Mind,* and what Daniel Pink calls "L-Directed Thinking" in his book *A Whole New Mind.* Anthropologists are not afraid to draw on their own instincts when developing hypotheses about the emotional underpinnings of observed human behavior.

4 **Anthropologists seek out epiphanies through a sense of "Vuja De."**
Everyone knows that feeling of déjà vu, a strong sense that you have seen or experienced something before, even if you never really have. Vuja De is the opposite—a sense of seeing something for the first time, even if you have actually witnessed it many times before. I first heard the expression from my friend Bob Sutton, a professor at Stanford, though I've also been told that it traces its origin to the comic George Carlin. Applying the principle of Vuja De, Anthropologists have the ability to "see" what's always been there but has gone unnoticed—what others have failed to see or comprehend because they stopped looking too soon.

5 **Anthropologists keep "bug lists" or "idea wallets."**
Anthropologists work a little like novelists or stand-up comics. They consider their everyday experiences to be good potential material, and write down bits and pieces that surprise them, especially things that seem broken. A bug list focuses on the negative—the things that bug you—while idea wallets contain both innovative concepts worth emulating and problems that need solving. Whether the idea wallet lives electronically in your PDA or is simply a low-tech index card in your back pocket, it can sharpen your powers of observation and your skill as an Anthropologist.

6 **Anthropologists are willing to search for clues in the trash bin.**
The Anthropologist looks for insights where they are least expected—before customers arrive, after they leave, even in the

garbage, if that's where learning is to be found. They look beyond the obvious, and seek inspiration in unusual places.

Over the years, IDEO has developed dozens of tools for Anthropologists. We've documented fifty-one of them in a set of action-oriented cards called the *Methods Deck*. The interrelated methodologies are organized into the four categories of "Ask," "Watch," "Learn," and "Try." But our enthusiasm for anthropology began with observations. We do extensive fieldwork to begin a project, to move it along, and to breathe life into a team when a project slows down. The process is remarkably similar to that followed by an inquisitive scientist or ethnographer. We watch human behavior in people's native habitat. We track customers or would-be customers as they interact with a product or service.

"The way to do fieldwork," Mead said, "is never to come up for air until it is all over."

When we go out in the field for inspiration, we try to observe with fresh eyes. Adopting a Zen-like "beginner's mind" is easier said than done, of course. But doing so makes a world of difference in gathering fresh observations. Margaret Mead is a familiar example of the archetypal anthropologist, studying cultures of the South Pacific in a series of books that challenged stereotypes about the imaginations of children and the limitations of so-called primitive societies. Mead believed you had to be there, you had to observe firsthand. "The way to do fieldwork," she said, "is never to come up for air until it is all over." Great minds through the ages have urged this technique. Charles Darwin, for example, was a born observer. He began by studying the faces of his own children and included photos of infants expressing their discomfort through crying in his book *The Expression of Emotion in Man and Animals*. More famously, Darwin joined the crew of the HMS *Beagle* as the ship's naturalist for two years of remarkable observations that helped inspire his classic *On the Origin of Species*.

Observers talk with people others have ignored and travel to faraway worlds. They subscribe to Albert von Szent-Györgyi's belief that discovery "consists of seeing what everyone else has seen and thinking what no one has thought." The Anthropologist humanizes the

The Anthropologist can draw on a wide variety of tools and techniques for gaining new insights.

scientific method and applies it to the business world. But seeing with fresh eyes may be one of the hardest parts of the innovation process. You have to put aside your experience and preconceived notions. You have to drop your skepticism and tap into a childlike curiosity and open-mindedness. Without that sense of wonder and discovery, you're likely to be blind to the opportunities right before your eyes.

History tells us that routine often blinds us to the truths that have been before us all the time. Until Jane Goodall applied her rare combination of patience and bravery to the study of chimpanzees, no one seemed to notice how much those clever primates share our ability to make tools, kiss, tickle, hold hands, and even, yes, pat one another on the back. The truth was there all along, waiting for us to discover it.

We can't all be Jane Goodall (or Margaret Mead, for that matter). Nor, in the corporate world, do we need to be. But approaching field observations with a spirit of curiosity can make all the difference in the world in identifying new opportunities or solutions to existing problems. So what makes a gifted Anthropologist? Patrice Martin, a bright young IDEOer with a degree in industrial design from the University of Michigan, has found her true calling as a human factors specialist. Patrice has an uncanny knack for getting people to talk about themselves. She looks even younger than her twenty-seven years and has a bubbling enthusiasm that's contagious. She might have been a star newspaper reporter in another life, because she quickly gets at the essence of a problem.

Why is she such a good observer? She truly enjoys meeting and talking to people. She asks probing questions that encourage people to reveal themselves. She projects a nonthreatening image that says it's safe to talk. She seems to have an intuitive sense of how to mine stories that unearth epiphanies into human behavior. For instance, Patrice recently worked on a project to develop healthy snack foods. Our client arranged for a series of interviews with doctors and patients—a perfectly reasonable approach. But Patrice took a less structured tack. She got permission to hang out in several pharmacies and talk with customers. Patrice made the initial contact, offering people discount coupons to encourage them to chat about healthy snacks. The men and women she talked to in drugstores were all over the dietary map: A middle-aged man looking for an energy boost while his wife was on the South Beach Diet. An elderly woman trying to combine two health drinks to meet all her dietary needs. A college student into natural foods, overwhelmed by the complexity of nutritional labels. A woman recently diagnosed with diabetes, confused about what foods would be best for her.

Armed with discoveries from her fieldwork in the pharmacies, Patrice next journeyed to the homes of a dozen people to learn more about food-preparation and eating habits. Spending time with people on their home turf not only makes them more comfortable, it also gives the Anthropologist a chance to look beneath the surface. For example, one woman in Patrice's field observations seemed to be the perfect homemaker, a virtual Betty Crocker. When Patrice arrived, she smelled the tempting aroma of a chicken baking in the oven. A healthy looking green salad and steamed vegetables were already on the table. As usual, Patrice had brought a video camera along to preserve her findings, so she was capturing this domestic scene on tape. If Patrice had spent only a short time there, she would have come away with a distinct—though misleading—impression of the family's eating habits. A few minutes later, however, the woman's kids arrived home and expressed stunned amazement on camera—"Mom, you cooked!?"

Patrice laughed as she told the story. "Her cover was totally blown. Later, we found pizza boxes and frozen-snack containers in the recycling bin." Patrice wasn't trying to bust this homemaker's meal-preparation skills, just get at her family's true eating habits. She found it easier to get the real story when she spent quality time with them at home.

Patrice asked a busy soccer mom to create a food map of everything eaten during the day. The woman wrote down three square meals and a couple of healthy snacks. Just to double-check, Patrice asked her, "You didn't eat anything else?" Without further cajoling, she admitted to a chocolate bar or two. Good Anthropologists paint a fuller picture. We're not looking for perfection, just authenticity.

One thing Patrice taught us about her experiences in cultural anthropology is that "life isn't typical." She never asks general questions, like "What's your typical diet?" In the process of generalizing, human nature causes people to idealize, which defeats the purpose of the observation in the first place. On this project, for example, she asked people what they ate that morning and the morning before. Says Patrice: "It's amazing how often people will say, 'Well, today was unusual.'" Today is *always* a little unusual. Life is messier than it is in a marketing brochure.

Patrice was looking for people's journey. She handed out "emotional stickers" bearing evocative words like *guilty, healthy, satisfied, balanced,* and *stuffed* to stick on their food map for the day. The words were meant to help express how people's food choices actually made them feel. Above where they described their meals was a separate line to put in what they wished they'd eaten. She also asked them to plot their energy throughout the day. The process created a series of richly textured food journeys that conveyed an individual and emotional sense of what people eat and aspire to in their daily routines.

It's amazing how often people will say, "Well, today was unusual." Today is *always* a little unusual. Life is messier than it is in a marketing brochure.

So how can you bake up some fresh inspirations? Enthusiastic Anthropologists are the yeast, skilled and interested individuals who actively seek out authentic experiences to observe. Trying to understand real eating patterns, Patrice wasn't satisfied just inviting people in for interviews. She sought out consumers where they shopped, and brought her camera and open-minded curiosity to people's homes. Patrice pushed to make her food maps more than just dry charts and statistics. They included emotional descriptors and a list of the foods people wished they had eaten. Her charts added a deeper human dimension to learning about the role of food in people's lives.

When you seek out field observations, remember: The more emotional breadth you gather, the better. The more human needs and desires you unearth for your experiential map, the more likely it is that they will lead you to promising new opportunities.

Human Extremes

Anthropologists have a knack for *not* falling into routines. There's a freshness to how they collect observations and dig up new insights. You've probably heard of "human factors," a technical term for the social science of observing people to gain an edge. But the term can be misleading, as it sounds slightly passive or academic. Human factors enthusiasts are highly proactive. They seek out the touch points of a situation—the key opportunities that have been overlooked or misunderstood.

If you want fresh and insightful observations, you have to be innovative about where and how you collect those observations. For instance, let's say you want to gain insight into improving a patient's experience in a busy hospital. Ask the doctors or nurses? Talk to lots of patients? Circulate a thoughtfully prepared survey form? All of these approaches sound reasonable, but IDEOer Roshi Gvcchi opted for a more radical technique. She calls it *Extreme HF*—short for "extreme human factors." Though not as wild as the extreme sports you see on ESPN, it's not for the faint of heart, either. Roshi, who has a background in film and new media, decided to bring a video camera right into the hospital room. With the permission of the patient and hospital staff, Roshi and her camera essentially moved in with a woman undergoing hip-replacement surgery. Roshi set up her video camera in

Classic anthropologists like Margaret Mead understood that fresh discoveries can be made by spending lots of time in the field.

the corner to run a few seconds every minute for forty-eight hours straight. To get the full experience firsthand, Roshi stayed in the room herself for two days, occasionally squeezing in a catnap in a reclining chair next to the bed. Following Margaret Mead's admonition, she didn't "come up for air until it was all over."

So what did her forty-eight-hour cinema verité capture?

The time-lapse video caught the ceaseless intrusions into a patient's room. Lights flipping on and off, doors opening, commotion in the hallway outside, nurses on their rounds. More than anything, it caught the astonishingly high number of people who entered the patient's room day or night. Roshi's flick was a bit like watching a vintage episode of *Candid Camera* or MTV's pioneering *Real World*. The images revealed how hospital staff bent various rules—like the number of family members allowed in the room at one time, or the visiting-hour restrictions—in their efforts to make the patient more comfortable. The video also demonstrated the impossibility of rest, let alone sleep, at some times of the day. Roshi edited down her forty-eight-hour time-lapse tape into an easily digestible five minutes—a powerful tool for understanding some of the problems and opportunities in a patient's room.

After seeing the video and talking to Roshi, I'm convinced that we're just scratching the surface for this novel technique. Digital video technologies have greatly advanced in the last few years, opening up many previously high-end capabilities to people without deep technical expertise. Though Roshi's media training helped her conceive, capture, and edit her time-lapse films, you don't need Steven Spielberg on your team to turn out evocative minivideos.

My advice is to pull out your video camera or find someone with a cinematographer's bent. What if you set up a camera to record the activity in a retail store? A lobby? A factory floor? Your offices? Not to spy on your staff, but to gain a better understanding of the ebbs and flows of your customers and your business. The next time you're looking for new discoveries, instead of holding a focus group, why not focus a lens on real customers, gaining insight into how people interact with your products, your services, your spaces. Body language says a lot. Imagine what you might learn if you could capture in images the circadian rhythms of your organization, the highs and lows of connecting to your customers. Imagine if you could use extreme human factors to gain new insights on what makes your customers tick.

Small Observations Pay

Picking up on the smallest nuances of your customers can offer tremendous opportunities. Recently, after giving a talk at the Food Marketing Institute conference in Chicago, I found myself surrounded by four large Polish guys who clearly had something they wanted to say. I was a little intimidated until one of them cracked a smile. It turned out that they all worked for a soft-drink company in Warsaw. They had cornered me because they wanted to tell me their own innovation success story. A few years back, they'd seen ABC's *Nightline* episode on "The Deep Dive" that illustrated IDEO's technique for learning from customers by doing field observations. After viewing the video together, they decided, "Maybe we could do that ourselves." So they set out for local train stations to look for clues about how they could sell more soft drinks to the captive audience of passengers waiting for the next train.

As they observed the crowds, they noticed a recurring pattern: In the minutes before trains arrived, people would stand on the platform, look over their shoulder at the drink kiosk, glance at their watch, and then scan the platform for the incoming train. A casual observer might have missed the clue. But these budding Anthropologists realized that passengers were torn between wanting something to drink and not wanting to miss their train.

So what did they do? They created prototype soft-drink displays boasting clocks so large that passengers could simultaneously watch the clock *and* the drink display. The result? Sales shot up in Warsaw train stations. The clocks reassured customers that they had time to buy a cold refreshment. That simple success made believers of these Poles. All inspired by watching a thirty-minute TV show.

We've been advocating field observations and quick prototyping for a long time. Sometimes a breakthrough is one small insight away. A simple telling observation—like the train passengers glancing from their watches to the soda kiosks—can make all the difference. Make patient observation and quick prototyping part of your recipe for innovation. You might be surprised by the results.

Interns & Intergenerational Waffles

At IDEO, the annual crop of summer interns is a continuous source of renewal for the firm. Some people think it's a form of organizational altruism that causes us to bring in more interns than we really need. Insiders know better. Not only does the intern program give us a leg up on recruiting decisions farther down the road, but it helps us stay fresh with a steady flow of ideas and irreverence.

For example, Michelle Lee, one of this year's new interns, recently spent several months watching grandparent-grandchild cooking experiences as part of her master's project for the Product Design program at Stanford. You may have heard of generation-skipping trusts, but this is generation-skipping in the kitchen. In a cultural anthropology program of her own design, Michelle noticed that the younger and older generations in some ways have more in common than the baby-boom generation in between. They live in the moment, not worrying about what they're doing next. They take time to savor the experience with all their senses, feeling the texture of the ingredients, smelling each new item, and liberally tasting the sweeter parts. Both young and old struggle with awkward or bulky items like heavy bowls and full bags of flour, and both seem extra attentive when their kitchen companion is handling a sharp knife.

Watching grandmas and grandkids sparked ideas for a line of intergenerational cooking products.

There are also times when the kids cover for their grandparents and vice versa. Grandparents have more knowledge, kids have sharper eyesight. Grandma knows what she's looking for, but her granddaughter can see it better. One cooking project Michelle watched while in

Anthropologist mode was a grandma making cookies with her four-year-old grandson and eight-year-old granddaughter. When it came time to read the recipe, Grandma had trouble with the fine print and the four-year-old had trouble with the words, so the eight-year-old stepped in to help out.

As her research continued, Michelle focused in on making waffles, a simple, rewarding process that all kids—and their grandparents—seemed to enjoy. The result is a line of product ideas she has for fun waffle-making. For example, all kids seem fascinated with breaking the eggs, but many struggle with the mechanics of getting that step just right. A fun, foolproof egg breaker that doesn't drop shells into the batter seems like a tool that these intergenerational cooking teams would buy in a minute. And that one idea may be just the tip of the iceberg for grandparent/grandchild products and services. The potential market seems huge, and grandparents seem willing and able to spend freely on such precious moments. So keep your eyes open for small insights in your field that can lead to new market opportunities. And in the meantime, never underestimate the power of an intern.

Kate's Seven Kid Secrets

We believe it's critical to observe and talk to kids. The freshness of their insights can't be found elsewhere. They see things adults skim right over. And there's no way to fudge their perspective. For one thing, you're not a kid anymore yourself. Your sense of childhood—and your view of the world—are filtered through layers of memory and shaped by the lens of adulthood. "I believe that kids have a certain kind of 'sixth sense' you don't find in most adults," claims Kate Burch, a designer who started her IDEO career in our Zero20 group—a team that gets its name from the age range of its favorite customers. And Kate reminds us that every generation's world is unique. "What it was like when I was eight is not even close to what it's like today. Kids today have a whole new set of opportunities—and a whole new set of pressures."

FIXED OPPORTUNITIES

If you take a close look at your world, you'll notice clever people playing the modern-day role of fix-it man. We've all seen the Post-it note with a helpful little instruction on top of the photocopier or the handwritten sign taped to the front of the reception desk. Perhaps you've been served by a resourceful salesperson or customer-service rep who doesn't do things by the book when the rules don't make sense. People can be ingenious and flexible when things don't work as advertised. They adapt technology and systems to fit their needs. At IDEO, we seek out these human touches in the field, these grassroots efforts by people to soften the sharp corners of the world, to offer a hand to help people along. They're signs that a product or service is incomplete. But they're also opportunities for future innovations.

Some opportunities are more obvious than others. To see how many exist in your world, try this exercise one day. Write down every fix you see at work, at home, or out on the town. Watch for things that have been duct-taped or bolted on. Look for add-on signs that explain what's broken or how a machine really works. You'll be surprised at how many you can spot. For example, enter most any urban taxicab and you're likely to see several little modifications added by drivers who spend their days and nights behind the wheel. And this quest for alterations and "fixes" is no idle exercise. Give it your serious attention and you'll have taken an important first step toward sensing the rough edges of many current offerings. You'll have a better understanding of why some products or services truly sing. And you'll learn to recognize when a product—or even a whole category—is crying out for improvement.

Kate has a natural, easy way of working with children. She makes it look effortless. What are her secrets? "It's all just common sense," she says. But from my experience, her gift is not that common. After reflecting for a bit, she comes up with one of the techniques she uses, and then another, and finally the ideas start to tumble out:

1 **Ask them about their shoes.**
Almost every kid has an opinion about their shoes. A big height difference is a barrier to communication, and a good Anthropologist wants to learn as much as possible. Get down on their level and talk.

2 **Offer something about yourself.**
Tell them a little about your day or your interests, especially something that shows your vulnerabilities; it will make you seem more human and help open new lines of communication.

3 **Ask them to invite their best friend along to talk.**
Even shy kids open up in the presence of a good friend, and they will provoke one another's storytelling. Sometimes, best friends will launch into an absorbing conversation on a subject and ignore you completely, which can be a remarkable thing when you're doing cultural anthropology.

4 **Remind them (only if it's true) that the project is "top secret."**
Even for kids who can't successfully keep a secret from their mother or their best friend, a little secrecy adds drama to the conversation and underlines the fact that you believe their ideas are important. We believe they're important, too.

5 **Ask for a house tour.**
Interview kids in their homes to gain fresh insights about the toys and things they like and dislike. Once they understand that Mom and Dad say it's OK, most young kids love to show you around. They'll jump from the macro tour of their home to the micro focus on the stuff in their room in five minutes or less. The house tour quickly becomes a window into the world of childhood.

6 **Ask kids what they would buy with ten dollars. Or a hundred.**
This question is an indirect but very effective way to find out what's hot and what's not. Ask a teenager about the latest gear

and you may just raise their defenses. But ask them what they'd spend a hundred dollars on and you'll get the real answer. What they'd buy is what's current, what's cool, what's top-of-mind for kids of that age.

7 **Make them laugh.**
Kids having fun have more to say. In a serious interview, they'll be on their best behavior, saying what they think you want to hear. But if you get them laughing, they're more likely to let you in on their real feelings, their real preferences, and give you the inside story on what it's like to be a twenty-first-century kid. They do less self-editing than the average adult, which is part of why interviewing kids can yield such insights. There's a lot you can learn from them.

Instant Observations

Even the most gifted Anthropologists sometimes lack the time or resources to do intensive observations. What can you accomplish when you're looking for a ready source of new ideas, fresh images, a sense of what's happening beyond your corner of the world?

At IDEO we believe in the quick provocation and information value of magazines and new books. We have an entire wall adjacent to my office filled with popular and edgy magazines for staffers to peruse, from *BusinessWeek* and *Fast Company* to *Dwell, Stuff,* and *Zoom.* They're not hidden away in some be-on-your-best-behavior style corporate library. They're placed in a big open room that's near one of the busiest thoroughfares in the firm. We believe that simply flipping through new magazines is a serious and productive practice for any organization interested in innovation. You might even find that it prompts your own publishing efforts. Our Consumer Experience Design group at IDEO (known internally as CxD) periodically produces booklets they call "Thought Bombs" to inspire the team. The Thought Bombs I've seen have been a fascinating collection of trends, concepts, and provocative ideas, mostly inspired by recent material from unusual print media sources.

To anyone who feels immune to the energy field around magazines, let me offer a suggestion. Drop by the Universal News and Café on Eighth Avenue in New York City. Imagine a generously sized bookstore, except that the more than 7,000 different titles reaching high up the walls are not books but glossy magazines. The intensely considered photography and arresting headlines of thousands of magazine covers in one place are so stimulating that they almost force you to deal with the store one section at a time. Even so, each of the store's categories has more titles than the total number of magazines you're likely to find at your local supermarket. I counted seven floor-to-ceiling rows—well over a hundred titles—just for science. One hundred and sixty auto magazines. More than a hundred and fifty on the subject of art and design. Separate international and foreign-language sections, each with dozens of titles in French, German, Italian, Spanish, as well as an entire row devoted to Africa. Universal News and Café is brimming with information, and the combined imagery of 7,000 titles has a certain magnetic quality that makes the store hard to leave. I'd venture to say that a few hours spent within its walls—there's a café to fuel you with ample food and drink—could tell you an awful lot about the trends and emerging vocabulary of just about any subject you care to research. There's no

A well-stocked magazine store can be a prolific source of fresh ideas and the latest trends.

skimping even on the hours: 5:00 A.M. to midnight gives you nineteen potential hours for intensive information retrieval every day.

What if you don't have a chance to drop by Eighth Avenue? Most major cities have a couple of stores similar in approach if not size to Universal News. Hollywood's World Book and News has 5,000 magazine titles. City Newsstand in Chicago tops out at around 6,000. In Miami's South Beach, there's the News Café. The major bookstores aren't bad, either—the biggest might even carry upward of a thousand magazine titles. Even if you don't spend an hour browsing—most of us have been conditioned *not* to—there's one piece of meta-learning you can pick up in the first five minutes: that there's more going on in the world than you can possibly keep up with. And *way* more magazines than you could possibly imagine. Spend some time looking at covers, flipping the pages, and, yes, even reading. You're likely to find some new ideas, not to mention a few new magazines you should subscribe to today.

First Look

Executives love to say that their company listens to its customers. In a world where there's always room for improvement, listening is mostly a good thing, but it's better at assessing the present than foreseeing the future. So even though detailed questionnaires can be really useful for assessing customer satisfaction, we don't really believe that the best breakthrough innovations come from asking customers.

Most customers are pretty good at comparing your current offerings with their current needs, and they're all in favor of something a little faster, cheaper, or easier to use. But they're not so good at helping you plan for new-to-the-world services, and they won't give you many clues to creating new business models. Asking them how to reinvent your service offering is a bit like asking someone on the street what NASA should do after it retires the space shuttle. Or even what product not currently on the market will change their lives in the next ten years. Those aren't the kinds of questions customers are well equipped to answer. There are just too many unknowns. Customers usually can't tell you how to create disruptive innovations.

But spend a day with them and watch what happens. Then you may actually start to get somewhere. If you're interested in making something new and better, you've got to watch people struggle and stumble. Take note of the people who pass by a shop because the entrance doesn't invite. Watch how would-be customers use your competitor's offering to see why they seem to prefer it. Some of the strongest clues to new opportunities can be found in the curious quirks and habits of people navigating their ever-changing world: how they respond to their environment, or exploit a novel situation, or adapt objects for their own use—often in ways the creators of those objects never anticipated. Some of these clever human adaptations are quite intentional, while others are almost unconscious. Jane Fulton Suri, IDEO's thought leader for our human factors work, calls these coping and response behaviors "Thoughtless Acts," and she has assembled a collection of her favorites into a book by the same name. Some of the insights you gain observing such thoughtless acts among your customers may be mere curiosities, but others may indicate a latent need that you could profitably serve. If you've got an open mind, these "acts" can spark your thinking—and maybe, just maybe, push you toward something new and authentic.

Practical Observations

Jane has helped me to see how anthropological fieldwork can be a disarmingly simple source of innovation ideas. Why do so few organizations practice this technique? Perhaps many just fail to act on the insights received. Good observations often *seem* simple in retrospect, but the truth is that it takes a certain discipline to step back from your routine and look at things with a fresh eye. I think organizations would send a lot more teams out into the field if they understood just how many business opportunities or cost savings simple observations can bring.

Part of what I've learned from Jane and other dedicated Anthropologists is that this work requires curiosity. How can you get better at it? Find a field that commands your interest. For me, it's travel. I do an awful lot of it, and by focusing on what works and what doesn't, I think I've become better at observations for a broad range of industries.

Not too long ago, for example, I literally stumbled onto an opportunity after a flight across the Atlantic. I was giving a talk outside of Paris, and like most overseas visitors to the City of Light, I flew into Charles de Gaulle Airport. My guidebook suggested heading into town via the urban train that connects the airport to the Paris Métro subway system. The train is superb, but it makes a pretty painful first impression. After buying a ticket for 7.50 euros, your first experience with the train station is to pass through the turnstiles on the way in. And that's where the trouble begins.

What fact did the architects—or, more likely, engineers—overlook? That nearly all passengers arriving on international flights would actually have *luggage*. The entrance did not seem to recognize the possibility of travelers carrying bulky suitcases, and the scene was so ridiculous that I stuck around for a while just watching people struggle. Not to take satisfaction in the suffering of my fellow travelers—for I had the same problems and sympathized with their plight—but just to observe human behavior and adaptive problem-solving.

As you attempt to enter the station, first you squeeze in toward a narrow turnstile. Once into that funnel-shaped space, you can't even

This turnstile near Charles de Gaulle Airport is easy to use—*unless you have luggage.*

carry *one* piece of luggage at your side, let alone the standard two. Since I travel light, with a twenty-two-inch black rolling carry-on bag and a briefcase piggybacked on top, I managed to squeeze through the first part of the station's unintentional obstacle course. But the classic three-pronged spinning stainless-steel turnstiles were like high hurdles for anyone with luggage. Those carrying two full suitcases were hard-pressed. While holding both of your bags at shoulder height—one in front of your body and one behind—you then have to slip a little purple ticket into a slot at the front of the gate and—worse yet—pick it back out of the forward slot at exactly the same time that you are spinning through the turnstile. Most passengers were dumbfounded at first, but they were motivated by the line backing up behind them and the desire to get to Paris. I saw "teamwork solutions" where husbands passed bags to wives on the other side. I watched individuals toss their bags over the barrier and then follow along. I witnessed balancing acts worthy of The Flying Wallendas. But I did not see a single person with two bags sail through easily on their first try.

Any good architect, engineer, designer, or machinist could come up with a host of simple solutions, but if and only if someone took the time to *notice* the problem in the first place. I only hung around for five minutes of field research and general entertainment, but presumably there are people who've been working near those turnstiles many hours every day for *years*. I'm sure most of these people must have witnessed this calamity hundreds of times. I suspect it's just considered to be "the way things are," something they'll fix in a decade, maybe when they expand the station or put in new electronic turnstiles. If only they'd first done a prototype—or even just considered that international travelers carry suitcases. Take the time to watch people or anticipate their needs, and I daresay they are less likely to get stuck.

Start Young

Anthropologists aren't valuable only for helping you understand today, they can also give you a glimpse of the future. For a look at *tomorrow's* mainstream markets, look at teenagers *today*.

A FASTER HORSE

A few years ago when IDEO was working with the Mayo Clinic on innovation, we had a small office in their Department of Medicine. I happened to visit the space one day and was struck by a Henry Ford quote the team had posted on the wall. "If I had asked my customers what they wanted," said the inventive Mr. Ford, "they'd have said a faster horse." Ford had a point. Don't expect customers to help you envision the future. Make that mistake and you're likely to get lots of suggestions for "faster horses."

Ford achieved many of his best breakthroughs in the early years of the twentieth century, but imagine you worked for a consumer electronics company that manufactured videocassette recorders in the first years of the twenty-first century. If you'd asked people what they wanted in a VCR, and let the question hang in the air awhile, they might eventually have suggested something like "super-fast rewind." You can imagine a customer saying, "When I am done watching a movie, I want to take it back to Blockbuster as soon as possible, so please give me faster rewinding!" How could you fail by listening to your customer? You might set out to create the fastest-rewinding VCR in the world. But just as you released your fancy new model, you would have been blown away by the arrival of the first DVD players—which, along with sporting superior image quality, sound, capacity, and improved reliability, require *no rewinding at all!* And as the pace of innovation accelerates, I hope everyone associated with the DVD format is preparing for subsequent innovations involving downloadable movies or video on demand, which will inevitably eclipse the same DVD players that had previously disrupted VCRs.

Of course, good companies still make a habit of listening to their customers. Just don't confuse that proven business practice with how you go about hunting up the next big breakthrough. That's not likely to come from asking people what needs improvement or fine-tuning. It's probably going to be something your customers haven't even thought of.

We've talked about extreme human factors. Central to these techniques is the idea that it pays to look at people who are a little different. People who love or hate a new product or service. People with opinions and biases who aren't afraid to express their feelings. Sound like a teenager you might know?

Teens try stuff constantly, check it out, and love it or chuck it. Prototyping at its very best. Kids ride the latest new technologies and fashions like the break at Waimea Bay. And when they do love something, their enthusiasm can help make it a hit.

Think of blogging, gaming, instant messaging, and MP3 file sharing. Teens helped drive all of these trends, and they're driving more as we speak. Pay attention to toys. They often inspire products that later captivate adults.

Kids are no strangers to IDEO. Indeed, our "lookout" space perched over San Francisco Bay with its racks of fun reading material and ever-shifting group tables sometimes feels a bit like a kindergarten classroom. And the common area of what passes for our management offices has a cluttered array of interesting objects and a full set of video gaming options that some days makes it resemble a teenager's room. It seems like every other week we're inviting kids to play with new toys and educational products to see what connections they make.

Of course, the toy-development portion of IDEO's Zero20 group has tapped into "kid power" for years to test out its countless toy prototypes. And get this: Founder Brendan Boyle discovered almost by accident one day that he could get more kids to show up on time if he charged a minimal hourly fee for playing with the prototypes. Moms were happy to pay (it was cheaper than babysitting), and the fee somehow triggered a psychological response that made them arrive early so as not to miss any of the valuable session.

Why do we watch and try to learn from kids and teens? They just soak up novel ideas, whereas grown-ups often spend a lot of time pushing back, telling you why it won't work. Text messaging, for instance, isn't necessarily the most efficient communication medium. But it spoke to teens' insatiable need to gossip and chat, and it wasn't long before adults lumbered on board too.

The Anthropologist has to start somewhere, and I can't think of a better place to begin than with the young. Whatever you do, in whatever industry you find yourself, make sure you watch and talk to teens and kids. We all know children make us younger in spirit. They can also help you see what's next.

CHAPTER 2
The Experimenter

I have not failed. I have merely found ten thousand
ways that won't work. —THOMAS EDISON

The Experimenter may be the most classic role an innovator
plays. Great inventors come to mind when we think of experimenters,
men like da Vinci and Thomas Edison. But when it comes to innovation, Experimenters don't need to be geniuses. What Experimenters
share is a passion for hard work, a curious mind, and an openness to
serendipity. Like Edison, they strive for inspiration but never shy away
from perspiration. We celebrate the Wright Brothers' success at Kitty
Hawk, but we often overlook the fact that they tested more than 200
wing shapes and risked their lives crashing seven different flying
machines in pursuit of a practical airplane. Few people stop to consider where the name for the ubiquitous spray lubricant WD-40 came
from, but it refers to the thirty-nine failed experiments in coming up
with the perfect water-displacement formula before the company finally
achieved success. And British entrepreneur James Dyson reports that
he built 5,127 unsuccessful prototypes of his cyclone vacuum before
he hit on the design that made him a billionaire.

Experimenters love to play, to try different ideas and approaches.
They put roller skates on the scientific method. They make sure everything's faster, less expensive, and hopefully more fun. Speed is an Experimenter's best friend. Experimenters embrace little failures at the early
stages to avoid big mistakes later on. They work with teams of all shapes
and sizes. They invite in colleagues, partners, customers, investors, even
kids to try out their works-in-process—all the possible stakeholders
who might have insights that could make the prototype better.

Who exactly is an Experimenter? At IDEO, we think it's someone who makes ideas tangible—dashing off sketches, cobbling together creations of duct tape and foam core, shooting quick videos to give personality and shape to a new service concept.

Experimenting in our world typically means prototyping, and prototyping is central to the IDEO tool set, as essential as a hammer is for a carpenter. Without the discipline of prototyping, we couldn't put flesh and bones on many of our new ideas. We've been prototyping so long that it comes naturally. That said, in the last few years we've learned a few more things about the art of prototyping.

First, you can prototype just about everything. Today, we prototype services as well as new products. Virtually every step along the ideation path can be prototyped—not just at the development stage, but also marketing, distribution, even sales. We've also learned not to be precious about prototyping. There was a time when we made a lot of beautiful prototypes. But prototypes don't have to be fashioned in machine shops or by designers. We cycle through prototypes, and our first prototypes can be pretty darned crude.

In the last few years, we've opened up the range of what we consider to be a prototype. Take proposals. We prototype them too. Recently, a major American professional sports league asked us for a proposal. We wrote up a standard document and got nowhere. But Experimenters recognize that the best time to try something really new (and risky) is when you have nothing to lose. After our initial effort was rejected—or simply ignored—one of IDEO's market-savvy Experimenters reminded us that a dry, colorless document seldom goes beyond its initial recipient, whereas a lively piece of digital content can go "viral" as it spreads like wildfire. We thought about how low-resolution video clips—usually some form of parody of current events—seem to race through a corporate grapevine. Although we had never used a video in this context before, everyone was game to give it a try. We created a simple, funny thirty-second video (accompanied by a one-page proposal) for the national sports league, capturing our own enthusiasm and the energy of the game. The video didn't take long to shoot and was so low-resolution that it could easily be e-mailed around. It turned out to be the perfect icebreaker. Just as we had hoped, the first league executive forwarded it to some of his colleagues, and they in turn forwarded it again. Within about ninety minutes, our fun little video reached the desk of the league commissioner. The low-cost experiment paid off, and we are currently involved on a collaborative project that we hope will score for the league.

Brendan Boyle and his Zero20 gang are natural Experimenters for the simple reason that it's central to the group's business. Brendan's team develops a wide range of products and services for kids and teens, prototyping hundreds of new ideas every year. And if the sheer quantity of prototyping has taught the team anything, it is this: Celebrate the process, not the tool. Sometimes you might shoot a quick video prototype. Sometimes a drawing will do the trick. Sometimes a rough storyboard will be enough to illustrate a multistep process. For physical objects, there are a host of digital prototyping tools like stereolithography or selective laser sintering. Other times all you may need is a simple piece of painted wood, enhanced with the wonders of computer graphics and video.

> Experimenters delight in how fast they take a concept from words to sketch, to model, and, yes, to a successful new offering.

Experimenters delight in how fast they take a concept from words to sketch, to model, and, yes, to a successful new offering. One morning Brendan's group was engaged in its daily brainstorm session when out popped an idea. What about a musical balance beam for kids to walk across? The idea was so simple and resonant—they immediately saw the parallels to the memorable scene of Tom Hanks dancing on the giant piano keyboard in *Big*—that they got right down to prototyping. They sawed off a two-by-six for the beam, painted the "musical sections" of the board in bright colors, and videotaped a young volunteer dancing sprightly on the beam. Then they polished up the video with some computer graphics, carefully matched musical tones to each footstep on the beam, and they were ready to go.

Total production time: a few hours making the physical prototype, and a day and a half producing the video. Within a month of the original brainstorm, a well-known toy giant bought the idea and started readying the product for market.

This ultra-quick prototype of a musical balance beam was enough to communicate and sell the concept to a major toy company.

That spirit is at the heart of an Experimenter. Not the tool, but pushing ideas into a more tangible, visual form as quickly as possible. What I'm suggesting is that you work to create an environment where it's OK to present less-polished prototypes. Serving up low-fidelity prototypes to your managers or clients requires a certain confidence. Crude prototypes require more courage than polished ones.

A good example of this happened in the first week of a project at IDEO Chicago for the surgical tool company Gyrus ENT. One of the early meetings brought us together with the company's medical advisory board to discuss desired features for a new nasal surgery device. We try to be on our best behavior in a room full of surgeons, knowing from experience that you need to treat them with respect. That said, our staff also understands that innovation relies on the free expression of even the most embryonic ideas. As the group described the not-yet-invented product, there was a lot of gesturing and hand-waving without much progress. Then one of our young engineers had a flash of inspiration and bolted from the room. Outside the conference room, seizing on the "found art" of materials lying around the office, he

Crude prototypes can increase the flow of new ideas for even the most sophisticated offerings.

picked up a whiteboard marker, a black plastic Kodak film canister, and an orange clothespin-like clip. He taped the canister to the whiteboard marker and attached the orange clip to the lid of the film canister. The result was an extremely crude model of the new surgical tool. Five minutes after his mysterious departure, he returned to the meeting and handed his kindergarten-quality prototype to a respected surgeon. He asked, "Are you thinking of something like *this*?" To which the surgeon replied, "Yes, something like THIS!"

That initial crude prototype got the project rolling. Amazingly, the sophisticated Diego Surgical System—today used in thousands of operations annually—traces its origins to that initial model. There's even a front rotating control ring reminiscent of the original marker cap. Had our young engineer subscribed to the "never show a half-baked idea to an important client" doctrine, he might not have taken that first leap of prototyping faith. He might have lost the opportunity to crystallize the insights of the surgeons. By taking a chance with a low-fidelity prototype, he was able to jump-start the project.

The lesson I learned from the Gyrus team was the value of lowering the bar for prototyping. Make it culturally acceptable to show off ideas at their rough, early stages and you'll see a whole lot more ideas. Incidentally, the first time I showed the image of this prototype to a business audience, a senior executive from Dallas asked me, "What if my company doesn't have the kind of creative people who can make this kind of prototype?" I looked at her, looked again at the prototype, and asked, "You're kidding, right? What part couldn't your people do? The Scotch tape?"

That's the beauty of lowering the bar. Literally anyone in the organization can float a trial balloon. I work with a group of talented designers who can whip up a drawing worthy of a magazine cover or render an object so realistically that you think it's a photograph. But they don't laugh

> I encourage the executives . . . to "squint" a little—to ignore the surface detail and just look at the overall shape of the idea.

at me when I go to the whiteboard and sketch an idea with my crude drawing skills. They just look for the idea inside.

This openness to low-fidelity prototypes may seem soft and intangible, but I believe that the social ecology of most organizations is extremely effective at communicating such subtle cultural clues. When a creative individual shows their boss—or even their colleagues—a good idea that's still a little rough around the edges, people pay close attention to what happens next. Does the organization build on the idea or ridicule it? Does management focus on the imperfections or the promise?

I encourage the executives of the companies we consult with to "squint" a little—to ignore the surface detail and just look at the overall shape of the idea. The informal communication system will spread the word quickly. If the "people who matter" in your organization learn to squint in this way, it will send a message to all the budding Experimenters that it's OK to try new things. In a culture of prototyping, you get previews of lots of ideas—even those not quite ready for prime time.

Extreme Prototyping

Remember I said you can prototype just about anything?

We had a team working to improve the life-changing experience of giving birth at a hospital. The problem at this hospital was that the maternity ward and postpartum floor were out of sync with each other. Though thousands of babies were born at the hospital each year, it was as if the two departments were not part of the same birthing continuum.

We decided we needed a firsthand experience of just what an expectant mother—and father—go through at the hospital. So we went undercover. We had a pregnant woman—who happened to be a member of the client's team—admitted to the maternity ward. Her husband wasn't available, so an IDEOer stood in. Amazingly, our ruse worked for large parts of the experience. The "couple" was able to go through the standard initial meetings with doctors and nurses—without anyone being the wiser. We skipped the actual birth process because the woman wasn't due for a month or so.

But that didn't really matter. We were seeking inspiration into how the transition from the maternity ward to postpartum floor could be improved. Our "undercover mother" was wheeled out of her maternity room on a gurney into the hallway and up several floors in the elevator. "Dad" cradled the baby—a plastic doll—in his arms.

Incredible as it sounds, the nurse in postpartum didn't seem to notice at first that the baby was a doll! The fake patient got away with it for a while, too, until the nurse pulled up her gown and looked at her with a puzzled expression. "Why do you have your pants on?"

After our client explained the elaborate experiment, the nurse sighed in relief. "That's good, because your baby looks really weird."

Our work led to a series of service prototypes that helped ease the transition for patients and improved communication and "handoffs" between the departments. Radical prototyping like this can do wonders for a project. Our client literally experienced the maternity ward and postpartum floor. She lay on the wobbly gurney and experienced what it felt like to be a patient. The maternity ward seemed in a big hurry to pass her on to postpartum. She experienced the separateness of the two departments, how they seemed to compete for resources, and how vulnerable and lost a new mother could feel in the transition.

Critics might say that good observers could see all this simply by watching and talking to new mothers. But innovation projects, especially those involving services and complex experiences, rarely take hold without collaboration and experimentation. If the client is literally on the gurney—feeling and seeing and thinking what a real mother goes through—that's worth a thousand interviews.

Experimenters engage the stakeholders in the prototyping process. They turn up genuinely useful observations. They start making the emotional connections necessary to bridge the gap between today's routine and tomorrow's innovation.

Implementation by Experimenting

Designing and implementing a service for multiple locations is fundamentally different from designing a product. You can't just churn out a new service in every city or town like identical cars rolling off a conveyor belt. The success of franchising in America might make one think this isn't so. It's easy to look at a superbly run restaurant franchise and think that, hey, if they could replicate the experience, why can't we? But here's what we've learned in the trenches about service innovations: They're ultimately about people and teams. They're about earning the respect and allegiance of the people who make or break a great new service.

Let me give you an example. At one point, we dove into a major collaboration with the flagship property of a major hospital network, generating a number of valuable service innovations at "headquarters." The traditional franchise model would be to stage a rollout of those new services at the satellite locations. But Peter Coughlan—head of IDEO's Transformation team that helps client companies grow a culture of innovation—points out, "Rollout sounds an awful lot like rollover." There's an inherent tension between systems and innovation. We believe in the importance of methodology and standards of service, but when you're dealing with entities with a level of independence—like the disparate offices of a large corporation or high-end hotel chains—you can't simply steamroll them with new ideas.

On this project and others, we've discovered that innovations developed at the flagship facility often require some translation to be successful at outlying locations. Peter often advises that we introduce the new services by asking the other locations to "put on an experiment." The reason has to do with how change occurs on a human and organizational level. Opposition can be fierce if an idea hasn't been invented or adapted willingly in the local system. "You want each location to be an R&D arm," says Peter. "If they're facing a new problem, you want them to come up with novel solutions."

Pressuring people to adopt outside approaches can stir up hostility and resistance. Instead, we invite the locations to prototype key concepts. We prototype two or more approaches for each solution, to make it abundantly clear that there's no single solution. That lets the local staff experience "what's in it for us." They can adapt a basic idea—say, a superior staff shift change—to the nuances of their local team. Most important, they're given the time and space to make the new changes their own. There's a certain honesty in making them integral to the prototyping process. We call it "implementation through experimentation." After all, they're the individuals who are going to have to make the new services a reality in what will to some degree always be a unique environment and set of circumstances.

Try making ongoing experimentation a part of your approach to creating services. Embrace experimentation and prototypes and say good-bye to the rollout. Be open to learning from successful experiments wherever they happen in your organization. You may be surprised. Not only will major initiatives you've undertaken be more likely to take hold, you're bound to encourage initiative in multiple locations that will keep your systemwide innovation momentum rolling.

Experimenting in Real Time

No company I've come into contact with lives the mantra of "implementation through experimentation" more than the Silicon Valley–based Tellme Networks. You may never have heard of Tellme, since they don't have much name recognition with consumers. But if you've used a voice-powered phone system, whether it's making a flight reservation on American Airlines or calling national directory assistance, you've probably used their intelligent voice-recognition software. I visited Tellme last year with Don Norman, an expert in user-friendly design. Three days earlier, Don had tried out their software by booking a flight to Los Angeles on American's voice-recognition telephone reservation system, and mentioned to a Tellme executive that the system failed to recognize "L.A." as the almost universal verbal abbreviation for Los Angeles. When we met with Tellme president Mike McCue, Don said, "I was very impressed with how well your system worked, except for that little hiccup about recognizing 'L.A.'" "Oh, that?" asked McCue with a surprised tone. "We fixed that already." I couldn't believe it. Could a gigantic real-time system, operating live in the national telecommunications network, have software written, tested, and released in three days?

"That's nothing," Tellme's VP of Caller Experience, Gary Clayton, told me later. "We can be much faster than that." He then told the story of a Silicon Valley dinner with executives from UPS while Tellme was wooing them as a client. Over prime ribs at the Sundance Mining Company, one of the UPS execs mentioned that they had rebranded the company recently and that they were no longer the United Parcel Service—just "UPS." "Your directory assistance system still says 'United Parcel Service,'" she noted. "It may not seem like a big thing, but it's important to us." Gary got up a moment later to make a phone call. Before they were finished with dessert, he handed the UPS exec a mobile phone. "Try it again," he suggested. "I think you'll like what you hear." Sometime between the main course and dessert, Tellme's software wizards had done a quick software update and patched it into their live system, getting the UPS name just the way they wanted it on Tellme's network services. It will come as no surprise that UPS is now a Tellme client.

The lesson of this story applies to all kinds of companies—from finance to manufacturing and retail. If experimenting is part of your culture, you can respond in hours or days, changing your offerings to meet market shifts and customer demands. Quick reflexes and fast turnaround can be part of what sets you apart from the pack.

Flushing Away Mistakes

"Fail often, to succeed sooner" is an old IDEO axiom. It's rooted in our philosophy of rapid experimentation. When your culture embraces the notion of lots of quick prototypes, you'll make lots of little mistakes that are really critical steps on the road to success.

Unfortunately, some people and organizations have been beaten up so many times over little mistakes that they have developed a fear of failure. And it's self-defeating. Fear actually makes failure more likely and experimentation nigh impossible.

So what can you do? A nationwide group called the Positive Coaching Alliance—started a few years ago at Stanford University—has learned that one big reason some kids don't enjoy sports is a fear of making mistakes. To combat the natural fear of failure, the Alliance advocates introducing what it calls a mistake ritual.

Think of it as a *success* ritual. The goal is to clear away mistakes to make room for success. Positive Coaching Alliance board member Dr. Ken Ravizza recently had a chance to put the ritual into practice at Cal State Fullerton. The school's baseball team had a losing record. Ravizza, a renowned sports psychologist and professor of kinesiology, set out to change the way the Titans thought about mistakes. If players struck out, hit into a double play, or had any other kind of morale-zapping failure, they'd come back to the dugout and literally "flush away" the mistake with a palm-sized, realistic-looking (and -sounding) toy toilet. At bat, they'd carry an image of the mini toilet in their minds. After a bad swing, they'd step out of the batter's box and mentally "flush" to clear their mind.

They had a group ritual too. After a tough loss, they'd circle around, strip off their jerseys, and toss them on the floor—to expunge the game. They even let go of that most American of pastimes, blaming the umpire. After a horrendous call, the new Titans would turn to the ump and thank him for calling a strike.

Suddenly, the Titans started winning. The team "flushed away" its 15–16 mediocre start and went on a tear, winning thirty-two and losing just six, and then, incredibly, rising out of the tough losers' bracket at the College World Series to win the national championship.

Could you come up with a symbolic way of letting go of mistakes at your company, or within your division or team? It can't hurt, and it just might turn your team into a winner, too.

Paper-Thin Prototyping

Many of us think of prototypes and innovations as massive, coordinated efforts, but I'm constantly amazed and encouraged at how little it takes for a good prototype to work. Sometimes the secret lies in figuring out how to address a single question—how, for instance, to make room for your product or service in a customer's already crowded life. For example, remember when the first wave of large forty-two-inch flat-panel TVs came out? Falling prices by themselves weren't enough to interest many families in buying the first wave of product. Retailers faced yet another hurdle. Large flat-panel displays eat up *lots* of wall space.

Consumer electronics marketers tell me that the physical differences between flat panels and old TVs alter the family decision-making model. They say old TVs fall into the "technology" category that was often the husband's domain, while slim new flat-panel displays fall into interior design, where the female decision-maker typically rules the roost. The geometry is so different that many people have a hard time imagining where—or whether—the new TVs would fit in their homes.

Mary Doan, head of marketing and advertising for The Good Guys electronics retailer, told me a great story about an experiment she came up with to overcome this barrier. During a trip to New York, she became intrigued by the cool fold-out "Z-card" maps she saw there, and wondered whether some new twist on the elaborate fold-outs could help her store sell flat-panel TVs.

That inspiration led to a clever Good Guys fold-out advertising piece, small enough to fit into a newspaper or magazine, that unfolded into an actual-size image of a forty-two-inch flat-panel TV—though, of course, this full-size prototype formed a *very* flat panel. When I saw the ad, I imagined thousands of households with the paper-thin flat-panel TV taped up on the living room wall, and someone saying, "See, honey, it could go right here." In fact, that's pretty much what happened. Sales of the flat-panel TVs bumped up the following month at The Good Guys, and one store manager reports that half a dozen customers came in one day saying they already had the paper version of the new TV taped to their wall.

The simple paper prototype was just enough to create a tangible vision and spark demand among people previously uncertain whether they had room for the new technology. My favorite story from Mary was about a single dad in Venice Beach who came home from work one November day to discover that his kids had taped the forty-two-inch ad up in their family room. "Dad, that's what we want for Christmas this year." And Dad reports that he was no match for the power of a paper-thin prototype.

How might you make it easy for potential customers to envision and prototype the idea of using your products or services? Could you use an inexpensive prototype to chip away at whatever is holding them back from becoming your loyal new customer?

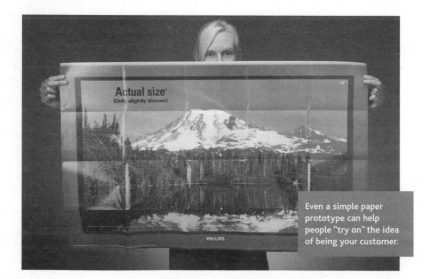

Even a simple paper prototype can help people "try on" the idea of being your customer.

Multiple Prototypes

Experimenters believe more is almost always better when it comes to prototypes. One prototype is like having a single rabbit: It has some value, but two can be more interesting, and can start you down the path to more and more. The trouble with a lone prototype is that if you show someone your one-and-only bright proposal and ask expectantly, "What do you think?" their answer is muddied by what they think about *you*. If they're a friend, they'll likely shower you with encouragement regardless of the idea's merit. But suggest an idea to your nemesis, and you're bound to be met by a withering "I don't get it!?!" At IDEO, we always try to present more than one prototype to guard against such fruitless responses. Battle-hardened Experimenters know that a variety of options makes possible a much more frank and positive discussion about the pros and cons of a prospective idea.

Here's an example from outside the business world that may have a familiar ring. After dinner one evening, my wife says, "Honey, I bought a new outfit today," and disappears into the next room to try it on. A few minutes later, she returns in the new dress. "Well, what do you think?" Of course, it's a charged question. You've gotta love that dress, right? It's still a prototype, since the sales tags are still on it, and it's still fully returnable, but that's not the point. She's asking about the dress—with her in it. She picked it out, tried it on, paid for it, and brought it home. I've gotta like that dress, because she's already committed to that choice. With only one option on the table, I'm either on her side or I'm not.

Experimenters recognize the value of introducing multiple prototypes. They transform the situation. Imagine that instead of just seeing the one chosen dress at the end, you have a chance to see seven dresses earlier in the process, as your wife hauls them into the department store dressing room. On the way in, she pauses to ask what you think. This time, flipping through the options, you can pick out the same exact dress and say, "Honey, I don't think this one will look good on you." Explain why it's not the ideal if you like, but that's rarely necessary. You're not saying she might not look great in one of the other dresses. Why can you speak the truth? Because you're not trapped in an awkward situation where someone you care about has already put all their eggs in one basket. Likewise, it's seldom a good idea to put your boss (or your client) on the spot. Don't demand a love-it-or-hate-it answer. Customers are rarely in a take-it-or-leave-it situation with one option to choose from. They're accustomed to weighing the pros and cons of multiple offerings and expressing their preferences.

Offer as many prototypes as possible within the limits of your budget and schedule. You'll avoid some awkward conversations. You'll get more honest, genuine feedback. And you'll learn from each prototype so that the finished result can be smarter, better, and more successful than the prototypes that got you there.

Some readers will doubtless think, "Sure, we'd love to have more prototypes—but we can't afford a bunch of costly experiments." And that's exactly why you have to lower the bar, making prototypes quicker and cheaper than ever before.

Chunking Risk

The value of making little experiments is especially important in a business that entails service or an experience. At Brigham & Women's Hospital in Boston, we encountered a bottleneck that snarls many a large corporate office. At lunchtime, the elevators were sometimes so swamped that foot traffic slowed to a crawl. In most organizations, you might look upon this as an inconvenience or inefficiency. At a hospital, it's a real problem. Family and friends, not to mention doctors, had trouble getting to see patients. We quickly brainstormed possible solutions. While the main elevators were swamped, the service elevators, reserved for moving patients and equipment, were underused. What if they put a guard on the service elevator to make better use of it without sacrificing the hospital's ability to quickly transport patients? To test the hypothesis, Brigham assigned a guard to that task for a couple of days. In the meantime, the team tackled the problem from a different angle. What if they could encourage more people to use the stairs? Simply ordering the staff to use the stairs seemed unlikely to produce results. So what did the team do? They sponsored a stair-climbing contest.

"Did You Take the Stairs?" read a prominently placed colorful poster board. There was a list of nurses, doctors, and technicians with stickers to affix next to their name for every time they took the stairs. Guess what? An awful lot of nurses (and some docs) caught the fever and the lunchtime elevator crunch eased. The contest worked on several levels. It developed awareness about elevator usage. It gave team members a chance to do their part, to pull together. Equally important, it signaled that it's OK to run a little experiment. To take risks.

That's the heart of an Experimenter, someone who loves to prototype. London-based IDEO designer Alan South calls it "chunking risk." Breaking down seemingly large problems into miniature experiments to the point where—lo and behold—you've generated system change without even knowing it. The power is in making lots of little steps at the same time, building momentum and optimism, the sense that one or a combination of approaches will deliver the necessary improvements.

The next time you're facing a complex bottleneck, give it a try. "Chunking risk" works.

Breaking the Rules

Sacred cows, cardinal rules, call them what you may. There comes a day when Experimenters need to break new ground by challenging some key assumptions. A few years back, the Minneapolis-based ad agency Fallon McElligott was trying to create the next television ad for BMW. But instead of "playing by the rules" by filming the industry-standard thirty-second commercial and placing a giant media buy to run it repeatedly in all the major urban markets, BMW commissioned some of the world's best independent directors to film eight-minute dramas dubbed "The Hire." The films defied the very definition of a commercial by *never appearing as a paid TV ad*—debuting instead online at bmwfilms.com. No standard commercial. No media buy. Great marketing buzz as BMW got free press and TV coverage around the world and car aficionados started a de facto chain letter as they passed the URL to all of their friends. Freed from the small-print rules of engage-

bmwfilms.com was a daring experiment that succeeded in generating terrific marketing buzz for the carmaker.

ment for TV advertising—like the obligatory notice that you're viewing a "professional driver on a closed course"—BMW got a chance to show off its cars' high-performance capabilities. And it was an ideal match for the emerging multimedia capabilities of broadband Internet access. After each new BMW production was released, the IDEO network almost visibly slowed down for a few days as, one by one, what seemed like all 350 people in the firm watched a streaming video of the edgy films.

The movies did more than drive tremendous traffic to BMW's Web site—over 50 million people have seen the films online, and I've even seen them show up on in-flight entertainment systems. The campaign was all about experimentation. The auto firm took a chance and was rewarded with record sales for two years running.

Prototype Selling

Today, it's not enough just to dream up a great new product. You've also got to figure out how to sell the darn thing. That can be the most important prototype of all. In the coming years, companies will need to be increasingly innovative about how they seed and sell their products and services. There's no assurance that what worked yesterday will work tomorrow. The first wave of Web-based selling—from Dell's efficient direct-to-consumer sales, to Amazon.com's one-click ordering, to Charles Schwab's online trading—all offered their innovators an early lead, which some maintain to this day. And the emerging use of unpaid "buzz marketing agents" today is a controversial—but so far successful—new way of using local thought leaders and trendsetters to smooth the adoption of new products.

Consider the classic story of Tupperware. Earl Tupper fashioned the eponymous Tupperware at DuPont from the leftovers of the oil business—polyethylene slag. Tupper quickly figured out how to refine and manufacture the material into a variety of potentially useful shapes. Selling his innovation would prove much trickier. In 1947, plastic had a lousy public image. Though Tupper developed a line of kitchenware tasteful enough to one day end up in the Museum of Modern Art, sales were poor. To begin with, the invention for which Tupper received his patent, the airtight seal, wasn't so easy to demonstrate. Prospective customers could rarely generate the distinct "burp" that was audible proof of product.

By chance, a friend gave a woman named Brownie Wise a piece of Tupperware. Wise practiced her "burping" skills, and was impressed when she dropped a container and it didn't leak. Though Wise worked as a distributor for another company, she asked if she could have a go at selling Tupperware. Her remarkable idea? She came up with the novel approach of giving home parties, inviting an ever-widening circle of new devotees to demonstrate the wonders of the new product. The rest, as they say, is history. Wise was soon racking up more than a thousand dollars a week in sales. Earl Tupper promptly caught on, dumped his shops, and hired on more Brownie Wise demonstrators to sell Tupperware directly to consumers. Sales exploded.

Today, companies still face this essential question. What's the best way to spark sales and build early buzz? A lot of turf is already staked out, so it's critical to experiment, to find the right medium to deliver your distinctive message.

Video Prototyping

At IDEO, we now frequently find ourselves prototyping in the world of experiences rather than in creating discrete objects. We're moving toward a theater where we're designing and harmonizing complex environments on video that encompass technology, architectural elements, and, of course, people. Not long ago, for instance, we were asked to investigate concepts for a line of spas. There was a time when the approach for such a project might be to do a lot of industry research and deliver a bound report to the client.

Here's what we did. We interviewed women—and men—in their homes about the subject of beauty products and salons. We toured and researched spas, talked to a range of people who might be spa users, and sought out interesting industry experts. We found a gentleman in Connecticut with a salon that resembled a classy wine bar and picked his brain. Girlfriends would go together to get manicures, as if it were a social outing. After we'd culled insights from a number of interviews and research, we started developing a vision of how our new spa concept might look and act.

How did we prototype the proposed salon? We wrote a script for a video prototype. We began with simple storyboards and brought in a freelance writer to help polish the short script. Our designers roughed out a futuristic computer graphics background of the prospective spa. The format was pretty straightforward: interviews with an owner and his customers. We actually auditioned an actor for the part of the salon owner, but it came off like acting, so we turned to our video guru, Craig Syverson, who by chance had the right voice, hair, and attitude. Craig looked and sounded the part of a hip salon owner, and he spoke with authority about what the new spa concept did for his business and clientele.

Most folks watching the short video for the first time would think there really *was* a chain of these start-up spas. The whole video was a prototype, of course, built on our observations and interviews with real people. It gave a face and a voice to our client's aspirations. We played the video on a big-screen display for our client and burned them dozens of CDs.

The video had the advantage of speed and economy. Sure, we could have built a prototype salon, but the time and expense would have been enormous. Even then, it wouldn't necessarily have given us a sense of the experience of a salon. The video prototype was a quick and extremely visual way of expressing how a salon might look and feel.

On another project, for Vocera, a promising Silicon Valley start-up offering voice-powered campuswide wireless communication, our video prototype ended up serving multiple purposes. Not only did it help visualize a new-to-the-world product/service combination, but it also helped them win a product design award and—much more important— win over new investors.

Any company can do their own video prototyping, but the approach used when you're in the Experimenter role is noticeably different from the one you might use for the Anthropologist. In Experimenter mode, videography is not a data-gathering or input tool. It's all about the output of sketch-quality communication, with just enough fidelity to get the idea across. And whereas the Anthropologist turns on the camera without a specific agenda, just to see what they will discover, the Experimenter prototypes with a point of view and a list of ideas they want to communicate. For that reason, it often makes sense to write out a script, and even storyboard the scenes in your video prototype. Given the ever-shortening attention span of corporate audiences, we'd suggest you keep your homemade video under six minutes in length (under three is even better, if you're up to the challenge), which means a script of no more than about a thousand words. After you've captured your scenes, edit them ruthlessly until just your core idea is there and is clearly, concisely communicated even to someone who doesn't know all the gritty details of your work. Think about the best possible case, in which your video gets passed up the corporate ladder a few rungs or—better yet—gets shown to an "insider" contact from your distribution channel or customer base. There's no need to polish it (you can always do that in the next iteration, if necessary), but make the tone and content suitable for audiences beyond the immediate one. So if you aspire to be an Experimenter, add video to your prototyping toolbox and you'll be able to have a bigger influence on the innovation process.

Play Time

I've already touched a little on how teens and young adults can help inspire new products. More and more, IDEO and our clients are realizing how critical the insights of the younger generation are, and not only for products solely targeted at their age group. We've done a lot of work educating our clients about tweens—what the *New York Times Magazine* recently called "those 8-to-12-year-old sophisticates with wallets." We consider tweens an entirely separate group, recognizing the often conflicted desire of these youths to be both independent and close to their parents.

It's not enough just to read about the nature of these tweens and teens. Our Zero20 group brings young people into the experimenting phase in all sorts of ways. Kids of all ages test our toys and products, and as I mentioned earlier, we often go to their homes—with parental approval—to see how they really play and act.

You can learn a lot by simply letting kids express themselves. More than a few books, stories, and movies have understood this basic truth. Think of *Willy Wonka and the Chocolate Factory*, a classic children's fantasy film (updated in 2005 as *Charlie and the Chocolate Factory*) that's enlightening for those who believe in innovation. Willy unleashes the creative potential of his Alice in Wonderland–style chocolate factory by inviting kids to help stir up the brew.

Letting kids express themselves isn't just fooling around. Listening to your youngest customers can really *pay*. Consider Danisco, a Danish food giant that claims half of the world's ice creams contain at least some of its ingredients. Danisco traditionally develops new products and then invites kids in to taste the results. In late 2001, however, the company flipped that process on its head. Much like Willy Wonka, the firm ushered a troop of children into its Copenhagen offices, gave the kids access to a roomful of ice cream and wild ingredients, and asked them to dream up new frosty treats. Many of the ideas were pretty wild, including a cow-shaped ice cream. Then, in one pivotal moment, a kindergartner asked, "Why don't you make some frozen jelly on a stick?"

Danisco's food wizards jumped on the idea. Rather than using the usual gelatin formulas, Danisco's scientists hit upon a unique combination of natural bean products and stabilizers. Even so, making their special new frozen jelly required a delicate process of first heating and then quick-freezing the ingredients. Prototypes were explored by a tiny specialty ice cream maker in Italy, which helped add and perfect natural flavorings.

And the new jelly pop did something unexpected. More precisely, it *didn't* do something. It didn't drip. And so the marketers at Danisco introduced the pop as the "nondrip lolly" and quickly gained the attention of consumers and the European press. I love the idea of a giant food company being humble and smart enough to invite kids in to dream up sweets. Off-loading some of the prototyping to a tiny ice cream maker in Italy. Stumbling onto something even better than the original idea. And probably having a lot of fun in the process.

Sometimes kids can provide the catalyst for a successful experiment.

Let Them Hack

The fact is that kids often make something out of what parents or companies would otherwise discard.

SMS, or Short Message Service, for instance, has been around for more than a decade. SMS was crudely designed as a text messaging system built into mobiles, originally intended for internal maintenance purposes only. Mobile phone companies "knew" it couldn't possibly be sold as a service. Cryptic and cumbersome, SMS was used by network technicians to troubleshoot problems.

Then irrepressible hackers discovered SMS. They invented their own abbreviated code to punch out messages. Hip teens caught on, and soon "C U L8er" meant "See you later" and a whole new wireless idiom developed. Part of the appeal was that the grown-ups were completely clueless. Indeed, the mobile companies didn't even know how to charge for SMS at first because the technology wasn't part of their business models.

> So who were the Experimenters who prototyped this new service and made it into a vast multibillion-dollar international business? Hackers and teens.

The world's major cellular firms just knew that SMS would never go mainstream. Which, of course, helped make it a wild hit. Kids loved SMS like they once loved walkie-talkies. Lo and behold, SMS's popularity built like a wave, moving into mainstream markets and appealing to people of all ages.

So who were the Experimenters who prototyped this new service and made it into a vast multibillion-dollar international business? Hackers and teens.

Virgin Mobile hasn't missed this lesson. They've taken that same sensibility and moved it upstream to teens and twentysomethings. As recently reported in the *New York Times*, Virgin has made its young customers part of the development cycle. Some 2,000 Virgin Insiders were asked their opinions on white or red versions of the V7 Flasher phone. They rejected both and instead favored blue with silver interior trim. Virgin went with the youthful choice. Next, some Insiders got to play with prototype phones, and the company soon learned that the kids were more interested in uploading photos to their blogs than the picture album they'd spent so much time building into the phone. Out went the picture album and in went streamlined uploads.

Call them teens, tweens, or plain old kids. They're probably not on your payroll or part of your business. They may drive you nuts at times. But listen carefully. They just might spur you to come up with a product or service you never imagined.

Life as an Experiment

Treat life as one big experiment and you'll start building a framework for continuous learning. And having a learning organization is part and parcel of a culture of innovation. The Experimenter helps keep the organization fresh and is willing to take calculated risks. Trace the history of any great innovation and chances are you'll find the footprint of an Experimenter. Sure, a few lucky souls have an apple fall on their heads or get a flash of enlightenment sitting under a tree, but for the rest of us, experimentation is one of the best ways to push toward the next breakthrough. So don't wait at the starting line trying to figure out the whole race. Just get moving and start trying things out. Along the way, you might discover a new way to win.

CHAPTER 3
The Cross-Pollinator

Leave the beaten track occasionally and dive into
the woods. Every time you do so, you will be certain
to find something that you have never seen before.

—ALEXANDER GRAHAM BELL

There's magic in cross-pollination—and in the people who make it happen.

Cross-pollinators can create something new and better through the unexpected juxtaposition of seemingly unrelated ideas or concepts. They often innovate by discovering a clever solution in one context or industry, then translating it successfully to another. For example, it was a Cross-Pollinator who transplanted the idea of a piano keyboard from the musical world to create early manual typewriters in the business world, which of course evolved step by step into the electronic keyboards we all use today. And reinforced concrete was originally created by a French gardener trying to strengthen flowerpots, but civil engineers wholeheartedly adopted it to create colossal dams and highway systems, while architects extended the gardener's utilitarian concept to elegant structures from Fallingwater to the Sydney Opera House. Computer pioneers got the idea for IBM punch cards—and arguably even the digital computer itself—from a punch-card system for weaving complex fabric patterns on a silk loom. The concept of an escalator began life as a primitive Coney Island amusement ride that has since grown into a billion-dollar industry. And most Frisbee players don't know that the basic shape and even the name of that ubiquitous flying toy was adapted from the Frisbie Baking Company's metal pie tins, tossed by Ivy League college students a century ago.

Pie tins from the world of baking were cross-pollinated into the internationally popular Frisbee.

Curiosity and an open mind have sparked cross-pollination opportunities throughout history. Food pioneer Clarence Birdseye, for example, was on a Canadian fur-trading trip in 1915 when he noticed his Inuit guides laying out fish to freeze in the cold outdoors, where it stayed fresh for many months. Cross-pollinating that simple technique from a native outdoor culture to his modern indoor world, Birdseye created a frozen-food empire that still bears his name.

Orville and Wilbur Wright cross-pollinated materials and mechanisms from the emerging bicycle industry to build their first powered aircraft. Now, more than a hundred years later, cross-pollination between cycling and aviation routinely flows in the *opposite* direction, as high-performance aerospace materials like titanium and carbon fiber are adapted to lighten and strengthen cutting-edge bicycles. And in the history of innovation, one of the greatest Cross-Pollinators, and perhaps the quintessential "Renaissance man," was Leonardo da Vinci—painter, architect, engineer, mathematician, and philosopher—who blended his diverse talents into a prolific and remarkable legacy.

In the corporate world, you can usually spot people in Cross-Pollinator mode if you know what to look for. They're the project member who translates arcane technical jargon from the research lab into vivid insights everyone can understand. They're the traveler who ranges far and wide for business and pleasure, returning to share not just what they *saw* but also what they *learned*. They're the voracious reader devouring books, magazines, and online sources to keep themselves and the team abreast of popular trends and topics. Well rounded, they usually sport multiple interests that lend them the experience necessary to take an idea from one business challenge and apply it in a fresh context. They often write down their insights in order to increase the amount they can retain and pass on to others. They're dedicated note-takers, capturing insights in notebooks or electronic form. Cross-Pollinators have eclectic backgrounds and develop a distinctive point of view by combining multiple strengths and interests.

Cross-Pollinating Inside and Out

Most companies I spend time with talk about cross-pollinating across organizational lines and "blasting through the silos," though in practice many of them have trouble doing so. Consumer products giant Procter & Gamble seems to have recently reenergized the Cross-Pollinators on their team since the arrival of CEO A. G. Lafley. Not only have they built on clever ideas imported from outside the organization (from the now-ubiquitous Swiffer duster to the playful Spinbrush toothbrush), but they also have gotten better at cross-pollinating ideas among previously siloed groups around the company. For example, they combined a knowledge of safe whitening agents from the laundry business with their deep expertise in oral hygiene to create Crest Whitestrips for their Oral Care unit—now grossing over $200 million per year. They combined experience in water purification from their PUR division with anti-spotting know-how from Cascade dishwasher detergent to create a "spotless" car-washing system called Mr. Clean AutoDry. There are dozens of other examples—already on the market or still in the works—that nimbly combine technologies and insights across organizational lines. You can see the effect of Cross-Pollinators at P&G not only on the shelves of your local supermarket but also on the share price, which has doubled in recent years.

Cross-Pollinators stir up new ideas by exploring worlds that may at first glance seem to have little relevance to the problem at hand. Peter Coughlan and our Transformation team often spark new service directions for our clients by deliberately staging cross-pollination outings with them. He takes companies on field trips to visit and observe analogous operations far outside their own domain. One client, for instance, held the belief that their tradition-bound industry didn't leave much room for innovation, so we took them to visit an enterprising undertaker. The client was stunned and ultimately inspired to see how much innovation was sweeping this (excuse the pun) moribund industry—everything from giving the bereaved a view of virtual custom coffins on giant projection screens to transforming the ashes of a loved one into a beautiful diamond. On a project to help a hospital optimize usage of their 600 hospital beds, we took our clients to a tiny New England bed-and-breakfast, where we found that maids working in two-person teams to clean guest rooms had more fun on the job than solo cleaners and were able to cross-check each other's work. That insight sparked the idea of replacing individual cleaners in the hospital with highly efficient small teams. Similarly, another hospital streamlined the way it transported patients around their large campus in wheelchairs and gurneys based on insights picked up on a visit to a well-run taxicab dispatch office.

Seeds of Cross-Pollination

The history of IDEO is itself a story of cross-pollination. When I joined the fledgling IDEO some twenty years ago, our workspace was cluttered with machine tools, physical prototypes, and other artifacts of a product-based business. I never would have imagined that someday we would be working on "intangibles" like improving Kraft's supply chain or helping nurses at Kaiser Permanente execute a more efficient shift change. But over time, we learned to apply our "design thinking" approach from product-innovation programs to the world of services, experiences, and even cultures. We've strived to nurture the Cross-Pollinator role from the very beginning, trying to assemble the key elements that encourage the flowering of cross-pollination. I have listed seven of the "secret ingredients" in our recipe for cross-pollination below, but of course there's nothing so secret about them now. And all of them could be translated to any company in the world that wants to bump up its level of cross-pollination and is willing to give them a try:

1 Show-and-tell. Whenever IDEO groups get together, we enjoy a hearty show-and-tell. In the early days of the firm, that meant sharing fresh insights or new technologies during Monday-morning meetings, when the entire company sat on the floor of my brother's office. The firm has gotten a lot bigger since then (and David's office got a bit smaller), so show-and-tell happens either face-to-face within smaller design groups or electronically across the firm via e-mail or our intranet-based sharing systems. The IDEO Tech Box, a collection of hundreds of promising technologies for potential application to our work, is a systematic approach to collecting and sharing what we know. Show-and-tell is partly serendipity, often resulting from an accidental discovery or surprise, so it doesn't always relate to projects the firm is actively working on right now. But it is always about something either new to the world or newly reinvented, and is a source of continuous renewal built into the work practices of the organization.

2 Hire lots of people with diverse backgrounds. We've never looked at hiring as merely a process of addition or bringing in "more of the same." If the recruiting task were to hire "another engineer just like Chris," then the interview would be a simple matter of pattern recognition. We're more likely to sift through the wide variety of applicants looking for someone who will expand our talent pool or stretch the firm's capabilities.

3 Stir the pot with space. As we'll discuss in the Set Designer chapter, the company's physical workspace can be a powerful tool for advancing your strategic agenda. Grouping all your like-minded people into one floor or building makes sense if you

want to emphasize solidarity in one discipline, but at IDEO, we believe there's magic in cross-pollination, and we support that belief with our use of space. We create lots of multi-disciplinary project rooms and leave ample space for "accidental" or impromptu meetings among people from disparate groups. We even make our staircases broad so that people can literally "meet halfway."

> No matter where you're from or how patriotic you may be, I hope you're willing to concede that there are more new ideas outside your country than inside.

4   Cross cultures and geographies. IDEO favors a cultural melting pot, seasoned with a steady mix of international flavors. No matter where you're from or how patriotic you may be, I hope you're willing to concede that there are more new ideas outside your country than inside. Importing new insights is always valuable. I've lost track of how many nationalities are represented in our firm, but a few years ago, our Boston office—just for fun—raised a flag for every country represented on their team. Last time I visited, there were eighteen flags hung, a pretty robust tally for an office of forty. And a well blended international staff just seems to cross-pollinate naturally from other cultures.

5   Host a weekly "Know How" speakers series. Nearly every Thursday evening, a world-class thinker shows up to share their thoughts with us. Not only are their insights often fascinating (Malcolm Gladwell on snap judgments, Howard Rheingold on smart mobs, Jeff Hawkins on the workings of the human brain), but the shared buzz of many IDEO people seeing a speaker sets off a wave of discussions throughout the firm. Know How is a weekly burst of cross-pollination that keeps the thinking—and the conversations—continuously fresh.

6   Learn from visitors. My role at IDEO includes the chance to meet with a continuous stream of interesting people who

travel long distances to visit us each year. Most are prospective clients who typically spend a couple of hours telling us about their industry, their company, and their point of view. Over the years, I've participated in more than a thousand such meetings, and I think of it as a form of postgraduate education. After each visit, I feel a little more up-to-date and attuned to current trends—and, dare I say it, just a little bit wiser for the experience.

7 Seek out diverse projects. There's an old saying that a forty-year career is sometimes the same year repeated forty times. Not at IDEO, or at any other company with a culture of continuous learning. The broad range of our client work—spanning dozens of industries—means that we can cross-pollinate from one world to another.

There's no rocket science involved in building a greenhouse for cross-pollination. None of these individual elements is especially hard to do. But put them all together—along with a hundred tiny details that support the social ecology of the firm—and they represent a commitment to cross-pollination that yields benefits in everything from team morale to competitive advantage.

Crossing Ideas

Cross-Pollinators are more than good students. They're good teachers as well, helping to spread knowledge and ideas. Ex-IDEOer Haydi Sowerwine and her husband, David, spent the first half of their lives in Silicon Valley, gathering knowledge and soaking up the culture. Now they've spent a decade transplanting IDEO-style design thinking to rural Nepal. Based in Kathmandu, their company, EcoSystems, has built dozens of gondola-like wire bridges over dangerous rivers in Nepal (at a fraction of the cost of suspension bridges), helping thousands of children get to school and villagers get to market

THE CROSS-POLLINATOR | 75

safely. Haydi and David's work recently earned them a Tech Museum Award for technology benefiting humanity, and they continue to expand their influence.

Cross-Pollinators retain the childlike ability to see patterns others don't, and to spot key differences. But they've also honed the very adult skill of applying those subtle differences in new contexts. They often think in metaphors, enabling them to see relationships and connections that others miss. They act as matchmakers, creating unusual combinations that often spark innovative hybrids. Cross-Pollinators frequently approach problems from unusual angles. They sometimes make a practice of "doing without"—tackling a problem by considering solutions without some key element popularly considered standard or essential.

Both the past and the future are great sources of ideas for a Cross-Pollinator, who revels in looking beyond their present challenges. Students of history, they search for concepts that may have been ahead of their time or may be ready for a revival. Conversely, they look for tertile ground in science-fiction stories, open to the possibility that imagined futures might provide a business opportunity today.

> I've spent a lot of time with T-shaped people, and one thing I've learned is never to leap to conclusions about them.

At IDEO, we've found that some of our most valuable Cross-Pollinators are what we call "T-shaped" individuals. That is, they enjoy a breadth of knowledge in many fields, but they also have depth in at least one area of expertise. I've spent a lot of time with T-shaped people, and one thing I've learned is never to leap to conclusions about them. It's tempting when you hear one salient fact about a person to start making assumptions, but with a T-shaped person, you are likely to be surprised by what you find out next. In the end, they defy simple categorization, but don't let that bother you. If you're looking for cross-pollination, gather some T-shaped people for your team.

IDEO is full of T-shaped people, and here's a quick glimpse to illustrate what I mean.

THE T-SHAPED PERSON

Empathy across Disciplines

(Coupled with)

T-shaped people are deep in at least one field while knowledgeable in many.

Deep Knowledge in Specific Areas

○ Kristian Simsarian at IDEO San Francisco has a bachelor's degree in computer science, studied robot perception, and did a stint at Edinburgh University's noted Artificial Intelligence Department, where he created his own course of study, spanning ethnography and engineering. He moved to Sweden to pursue a Ph.D., while building digital narrative tools for teachers in Europe. Kristian started his own improvisation group in a Stockholm studio space, and he continues to use improv as an ideation tool at IDEO today. He's a one-man multidisciplinary team, with deep understanding of human/computer interaction, broad interests in the learning process, and fluency in the freewheeling language of improvisation.

○ Sabine Voegler combines the cultural influences of her German father, her Brazilian mother, and her California experiences. Having lived on three continents, Sabine is currently at IDEO Munich, where she draws on her uniquely blended worldview to create remarkable customer experiences.

○ Owen Rogers, an avid IDEO Cross-Pollinator in our Consumer Experience Design group, describes himself as a former mechanic, stonemason, and disk jockey who "talked his way into" London's revered Royal College of Art. He has a deep expertise in managing large innovation and branding programs, but still maintains his interest in the world of auto mechanics, so he jumped at the chance to work with a client recently on high-end automotive tools.

○ IDEOer Kara Johnson, a quintessential T-shaped person, has deep expertise in materials science, with a master's from Stanford, a Ph.D. from Cambridge, and a successful book on the subject. Lest you think she's one-dimensional, however, she also has interests across a broad range of design fields, and she studied sculpture and pottery at Michigan's prestigious Cranbrook Academy. Kara worked on more than fifty projects during her first year at IDEO, spreading the word about new-material options we hadn't fully considered in the past. She has sparked new interest in materials across the firm and encouraged companies to consider alternative and sustainable materials.

Not all Cross-Pollinators are as versatile or multifaceted, but the good ones can send shock waves through an organization as they bring in big ideas from the outside. And Cross-Pollinators don't need to be brilliant inventors or titans of industry. Even small, pointed insights can make a remarkable difference.

Innovation on a Shoestring

I wonder if innovators everywhere couldn't get some inspiration from the grassroots entrepreneurial efforts of Mohammed Bah Abba, a teacher with a business degree in impoverished northern Nigeria. Bah Abba wanted to help keep food from spoiling so quickly in the intense African heat, but he knew that a standard refrigerator was out of the question for many of his neighbors. Bah Abba cross-pollinated from the past to help make a better future. Descended from a family of potters, he began adapting traditional clay pots and hit upon something remarkable. When he placed one pot within another, filling the space between with wet sand, the water in the sand evaporated toward the outer shell of the inner pot, cooling the vegetables inside. He spent two years perfecting his clay "fridge," learning to cover the pot with a damp cloth. His cooler required no energy, just periodic wetting of the sand to maintain the cycle of evaporation.

Eggplants that once spoiled in a few days lasted four times as long. African spinach was edible for almost a week instead of going bad in a day. Bah Abba put unemployed potters to work turning out thousands of his clay pots, at a cost of 30 cents per cooler. Today, the lives of thousands of Nigerian villagers have been improved by this brilliantly simple innovation.

There's a principle at work here that we would all do well to respect. Sometimes a lack of resources and tools can prove to be the spark that helps you to seek out and make new connections. It goes beyond the idea that "necessity is the mother of invention." Scarcity and tough constraints force you to break new ground because the "business as usual" path is simply not available. There's an essential truth behind the Silicon Valley legends about companies getting their start in a garage. Lacking money or staff, they had to be resourceful.

Amy Smith, an instructor at Massachusetts Institute of Technology, demonstrates how innovators can turn a resource constraint into an opportunity. How does Smith plunge students from a prestigious New England university into a scrappy mind-set for innovating on the cheap? For one week during the semester, her students have to scrimp by in Cambridge on $2 a day. Along with dealing with hunger, students quickly learn that most of all, they have to be creative to get by on so little. Smith's program has been a catalyst for promising new ideas like a low-cost kit to remove land mines in Zimbabwe; a water-testing rig made with a Playtex baby bottle that costs $20 instead of the traditional $600 setup; and charcoal fuel fashioned out of the inedible remnants of sugarcane.

Amy Smith's MIT program makes me wonder what opportunities we might be missing in the business world because we take resources for granted. To create something new, you may have to take something away. For example, MTV does what they call "deprivation studies," where they get their most frequent viewers to go "cold turkey" for thirty days of no MTV, just to see what clever alternatives they come up with. So try your own version of scarcity. Spend a day generating and communicating ideas without the use of technology. Pass an afternoon prototyping without conventional tools. Like poets working with meter and rhyme, great Cross-Pollinators seek out constraints. The next time your ideas seem stale, challenge a team to come up with something on the cheap. It can be a great innovation exercise.

Increase Your Fluency

Cross-Pollinators are like linguists, confident in the knowledge that the more languages they master, the easier it becomes to absorb the next one. That's one of the secrets of cross-pollination. Diverse and interesting project work can fuel the fire of a culture of innovation. Give your team greater variety and they will start seeing the outlines of new connections, making new leaps of imagination.

For instance, not long ago we were asked to help redesign the computer science building of a major university. The straightforward approach would be to benchmark lots of other computer science centers for inspiration. Abandoning that traditional model, we instead took our team to Pixar's animation studio, across the bay from IDEO's San Francisco office. Sure, Pixar had plenty of computers, but everything else was remarkably different from the standard university (or corporate) lab. The emphasis on collaborative technical and human resources. The distinct work clusters, or "neighborhoods." Even the great food. Pixar resembles a teeming, electronically charged urban village, a provocative contrast to a university laboratory.

Cross-pollinating works both ways. If universities can learn from a company like Pixar, what might companies learn from Stanford or Harvard? Plenty, I'd suggest, if they toured the campuses with an eye to ideas or concepts that might take root on their own business turf. An open mind is central to cross-pollinating. And the more receptive you are to diverse approaches, the more likely you are to come up with something valuable for your own company.

The Fosbury Flop

Cross-Pollinators sometimes tackle a problem by turning it around. Creativity guru Edward de Bono called it "lateral thinking"—looking at an issue from a completely different perspective. Sometimes you literally need to sneak up on an old problem from a new direction. Instead of heading straight at the challenge, you approach it, well, backward.

To me, one great visual metaphor for "backward" innovation was the high-jump technique dubbed the "Fosbury Flop." Back in the 1960s, a kid named Dick Fosbury was a mostly unremarkable track-and-field athlete at his high school in Medford, Oregon. Fosbury preferred the familiar jumping style called the "scissors," a move you sometimes see on the tennis court when a triumphant player jumps over the net sideways. The scissors jump works pretty well on the tennis court, but his coaches knew that in the high jump he'd be better off with the more efficient and then-popular "straddle"—also called the "belly roll"—in which you throw up your lead foot and follow it with leg, thigh, stomach, and head, swiveling up and around the bar, with trailing foot clearing last. Fosbury did as he was told but was no better than average at the straddle, never eclipsing 5'4". But during one track meet at the age of sixteen, he started scissoring again, going against the conventional wisdom that it would limit his success. Then something unexpected happened. Gradually, as the height was raised, "I started laying out more," he recounted, "and pretty soon I was flat on my back." Fosbury wasn't quite flopping yet, but he was going mostly backward—and clearing 5'10", higher than he'd ever gone before.

The summer after he graduated in 1965, Fosbury started doing his trademark "Flop"—bounding in long, powerful strides, then, at the last moment, twisting to turn his back parallel to the bar and leaping into an arched backward half-somersault—shoulders up, followed by knees, with both feet clearing the bar at the final moment, then landing upon his shoulders, faceup. That summer he flopped over a bar 6'7" high and won a national junior championship. In college, once more his coaches tried in vain to set him straight—but he could never

Dick Fosbury fashioned a radical new way to "flop" that won him an Olympic gold medal.

straddle more than 5′10." Fortunately, his coach gave up his plan to make "a triple jumper out of him."

Fosbury, of course, kept right on flopping—all the way to the Olympic Games. At the 1968 summer Olympics in Mexico City, all 80,000 spectators in the stadium seemed to go silent each time he jumped— backward. I remember watching the Olympics on TV with my father, who said, "Did you see *that*!? Watch this guy when he jumps again. It's the weirdest motion you ever saw." Like a lot of breakthroughs, the Fosbury Flop looked strange the first time you saw it. Really strange.

Experts said Fosbury would break his neck. Instead, he broke the American and Olympic records with a jump of 7′4¼″ and won gold. It took almost ten years for Fosbury's innovation to ripple through the ranks of elite athletes, but eventually the Fosbury Flop was fully adopted by every Olympic high jumper in the world. Incredibly, what Fosbury discovered in the mid-1960s—starting with a jumping style from other contexts and evolving it through enlightened trial and error—remains to this day the most efficient way to high-jump. His radical style permitted a faster run at the bar, with less deceleration than the straddle. Decades later, experts would publish elaborate biomechanical studies proving the Fosbury Flop's superior "angular momentum" and "somersault rotation."

Like a lot of break-throughs, the Fosbury Flop looked strange the first time you saw it. Really strange.

Looking back today, it's abundantly clear that the old straddlers were jumping with an outmoded technique. It took an independent thinker like Fosbury to fashion a genuinely new approach. But Fosbury didn't flop all at once. The breakthrough didn't come in the kind of "eureka" moment so popular in the mythology of invention. He experimented with a style widely considered flawed, adding his own twist, gradually refining his technique, never sure whether he was on a path to success or stumbling down a blind alley. As with many new business innovations, Fosbury was first told that his approach would fail miserably.

I can't think of a better moral for those interested in innovation. The next time someone tells you no one's done it that way before, or that it sounds like a crazy idea, ask them if they know the story of the Fosbury Flop.

Cross-Pollinators keep an open mind. They know that success can come from the most unlikely of all directions.

The Germ of a Seed

Cross-Pollination begins with people: individuals with restless curiosity and unusual backgrounds who expand your ability to tackle challenges. Some might find the résumés of our best Cross-Pollinators surprising or even eccentric. At IDEO, we've found a couple of great sources of Cross-Pollinators. Of course, we've always relied on a steady flow of fresh thinkers with eclectic experiences right out of college, individuals whose curiosity is still running strong. More recently, however, we're also tapping into the cross-pollination potential of "boomerang" staffers—talented people who worked with us for a while, have gone out and gotten broad experience elsewhere in business, and then come back.

Stanford product-design graduate Bob Adams is one such boomerang Cross-Pollinator. More than a decade ago, Bob spent two years working in India with Hewlett-Packard before joining an earlier incarnation of IDEO. After a couple of years, Bob left to pursue a career as a jazz bassist, getting the chance to play with such luminaries as Joe Pass, Richie Cole, and Stan Getz. Meanwhile, he did a long stint with a Silicon Valley think tank, exploring how musical controllers might improve the digital interface. He also made his own electronic instruments, and he taught at Stanford and briefly at the Royal College of Art in London. Somewhere along the way, Bob found the time to earn a master's in viticulture from the University of California at Davis, and discovered the energy and drive to buy a small farm in the Sacramento Valley. He calls himself a "weekend farmer," but the phrase doesn't do him justice. Bob mans the tractor and other heavy equipment to farm wheat, tomatoes, olives, peaches, and nectarines. His vineyard produces a respectable Zinfandel. He enjoys that rarest of twenty-first-century pleasures—the opportunity to sit down to an entire meal grown completely by himself.

Bob rejoined our San Francisco office a year or so ago, and he has already brought a thoughtful eye to new projects. Bob's passionate about sustainability, and seems well attuned to bringing more sustainable practices to business. More than just immersing himself in the subject and networking with some of the field's leading thinkers, Bob has experience running the ultimate sustainability model—a farm. He knows firsthand how hard it is to farm organically, and understands why farmers use pesticides and fertilizers. That grounding may help make his approach more successful in industry.

Every company could use some good Cross-Pollinators to enliven their culture and lend fresh perspective and experience to their endeavors. You might find their backgrounds surprising, but give them a chance to find some fertile soil and you won't regret it.

Found in Translation

To those who complain that there's nothing new in their industry, I say get on a plane and see the world. Traveling often and widely is one of the most effective ways to become a better Cross-Pollinator. Sometimes the most direct route to innovation is to look abroad and translate what you find.

One of my favorite shopping spots in Tokyo—a city of fascinating retail experiences—is a store called Mujirushi Ryohin, better known by its nickname, "Muji." The name translates roughly as "No brand. Good stuff." I think of the store as distinctively Japanese. It turns out, however, that the unique retail chain got its inspiration in America and is one of those stories of something being *gained* in translation. Back in the 1970s, when the Seiyu chain of low-cost department stores was trying to create a new "house brand," they sent a design team far afield looking for new ideas. One team went to the United States, searching for inspiration. During the trip, team member Kazuko Koike (who later became a celebrated art writer) dropped into an American supermarket searching for unusual beer cans to take back for a friend in Japan who collected them. She came across simple "generic" beer, which was part of a series of deliberately unbranded products, with stark black-and-white labeling, that were popular at the time.

Muji was able to take a simple idea from America and turn it into a stellar Japanese success. What ideas are out there waiting for you to tap into?

She liked the simple graphics and iconic no-frills design, and brought the idea back home to explore at Seiyu, which translated the American concept into a stylized, intentionally low-key Japanese design that was "the essence of generic." Seiyu's line of clothing and houseware products use simple materials like unpainted aluminum and minimalist packaging like unbleached paper, with a natural color palette limited to white and black, brown and gray. The resulting "no brand" Muji brand was so successful that, a few years later, Seiyu launched freestanding Mujirushi stores, starting with a location in Tokyo's fashionable Aoyama district. Today, Muji has almost 300 stores spread out from Sapporo to London, and sales of nearly a billion dollars.

That's the heart of cross-pollination. It's a rich source of inspiration for those willing to travel and imagine. Muji was able to take a simple

Simple Muji products reinforce their values of "No brand. Good stuff."

idea from America and turn it into a stellar Japanese success. What ideas are out there waiting for you to tap into? What might U.S. health care firms learn from extraordinary international models like the Aravind Eye Hospital in Madurai, India, a hospital that has done a million cataract surgeries at a cost of roughly $10 each? What popular regional foods, like *açaí* from Brazil or *edamame* from Japan, could be marketed successfully in large markets elsewhere? What concepts from a distant land or a foreign culture might you translate or shape or adapt in a way that makes them distinctively your own? So travel widely when you have the chance. Scour the world for ideas. Reinvent something native to Asia or Europe or the Americas. You just might have a hit.

Reverse Mentoring

The old adage "Mighty oaks from little acorns grow" may be true, but what do you do when your "acorn" days are far behind you? How do you continue to grow and flourish? Mentoring apprentices and protégés has been part of business as long as we've had crafts and professions. But when you've put a few growth rings under the bark, consider the flip side. Sometimes what managers really need is a mentor from a younger generation to inform and inspire.

The smartest folks I know have what my Stanford professor friend Bob Sutton calls an "attitude of wisdom": enough knowledge to sense when you're on course, enough humility to know when you need help navigating. In our experience at IDEO, we've found that it helps to be open to fresh approaches, even when all your experience suggests maintaining the traditional view. Reverse mentoring can help counter your company's natural tendency to be overreliant on its experience. Consider seeking out younger mentors to provide insights and initiative about what's happening in the world today.

IDEOer Chris Flink has been my de facto reverse mentor for the last few years, though we never really formalized the relationship. He's one of the first people I go to when I'm feeling behind the curve on some new trend that seems like it might be worth knowing about. For example, I noticed a couple of years ago that a lot of younger people around me don't wear watches anymore. So I asked Chris, "What's the story? You've got client appointments. Why no watch?" His response took me by surprise. "Tom, why do I need a watch? My mobile phone keeps perfect time. I don't have to manually adjust it for daylight savings, and when I go to new time zones it updates the time immediately." My first thought was *Wow, there's something that changed about the world when I wasn't looking.* And my second thought was *What if I were Timex? What's my business strategy for a future in which mainstream customers see no need for my product category?*

Reverse mentoring isn't widely practiced yet, though I know of a handful of companies that have realized that a fiftysomething executive could use some of the insights and enthusiasm of a twentysomething. My brother David learned the value of reverse mentoring a couple of decades ago when he found his spiritual home as a professor in Stanford's product-design program. David puts a lot of heart into his Stanford role, but he also benefits from his devotion to teaching. For one thing, he is showered with the ideas and enthusiasm of an unending stream of smart and motivated nineteen- to twenty-three-year-olds. And these bright young minds keep him informed and up-to-date in a way that those who spend their time exclusively in industry seldom are.

Long before Napster and Kazaa made front-page news in the nineties, my brother knew all about downloading music from the Web. Computer-based video editing showed up in his classes way before it went main-stream in the general population. Same for blogging, instant messaging, and a dozen other technologies embraced by the early adopters in his classrooms. David gets a window into all sorts of trends in fashion and music and video gaming, keeping him up-to-date with—sometimes ahead of—what's happening in the world. Nor is the knowledge he gleans focused entirely on culture and entertainment. David mentioned to me a couple of years ago that his students had shifted en masse from the atti-tude of "How can I get rich by launching a new product?" to "How can I introduce more of a social conscience in the world of business?" He saw it in the student population years before it really showed up in business discussions. So as part of the process of interacting with student groups, David learns not only what they *buy* but also what they *think*.

Could you benefit from a reverse mentor? Be one yourself? The best part of this cross-humanizing technique is that everyone gains. Con-sider opening a new line of communication, adopting an attitude that frees you to learn from the youngest members of your staff.

David calls it the eggs teaching the chickens.

The Gift of Giving

Giving may be the most counterintuitive and extreme form of cross-pollination. You're in business to make money, of course. But generos-ity can help you get there and be good for your company's karma. Generosity can be the quality that propels you above the crowd. As I wrote the final chapters of this book, major retailers like Niketown, Gap, the Discovery Channel stores, and others were finding that cus-tomers have a sincere wish to donate to a good cause—everything from Nike helping Lance Armstrong raise $33 million for cancer research with simple $1 yellow bracelets to Amazon.com generating $15 million in "1-Click donations" for victims of the catastrophic Asian tsunami.

The good karma of giving can be a surprisingly powerful, inspirational force. And you can take giving to a level you might not expect. Sometimes the most strategic branding move you can make is to give away what seems to be your crown jewels. One inspired Cross-Pollinator can make the difference.

Though you've probably never heard of Nils Bohlin, he had a lot to do with Volvo's success over the years. During the 1950s, Bohlin worked for Swedish aircraft company Svenska Aeroplan, where his specialty was ejector seats. In 1958, he became Volvo's first safety engineer. At the time, the two-point seat belt that stretched across your lap was considered state-of-the-art technology—and most cars in the United States had no seat belts at all. So Bohlin, who had spent years dreaming up ways to pop people out of airplanes, came up with a novel way to keep them in. He tossed out the less effective two-point belt and

Volvo inventor Nils Bohlin created the original three-point seat belt—*and then shared it with the world.*

ushered in the three-point shoulder belt, which remains the basis for nearly all automotive seat belts today.

By 1963, Bohlin's invention was standard on all Volvos. But to my mind the unique part of this story is Volvo's extraordinary and brave decision not to patent this remarkable invention to encourage the saving of lives. That pivotal choice and decades of leading-edge research and design have helped make the Volvo brand synonymous with safety. Bohlin went on to lead efforts that resulted in Volvo's highly acclaimed Side Impact Protection System. Today, more than forty years later, Volvo's slogan, "For Life," echoes the company's sincere commitment to making the safest vehicles in the world.

What might you give away that will give you an edge?

Emulating Nature

Those who practice cross-pollinating, perhaps more than any other persona, intuitively understand the role of serendipity and chance. By actively seeing and connecting with more ideas and people, the Cross-Pollinator becomes a bit like the unlikely bumblebee. Many have wondered how the bumblebee flies at all, with its bulky body and tiny, fragile-looking wings. But the bumblebee doesn't know that, so it goes on flying anyway. Perhaps the answer lies, as it does with so many things hard to comprehend, in the sum of the parts. And so it is with the Cross-Pollinator, a sometimes unsung role in the business world, the person who tirelessly spreads the seeds of innovation.

As you may have figured out, the Cross-Pollinator is in many ways a collection of personas—part Anthropologist, part Experimenter, part personas you have yet to meet. Every organization needs Cross-Pollinators. Maybe, like the bumblebee, you too are an unlikely hero. Do you have wide interests, a voracious curiosity, and an aptitude for learning and teaching? Are there others on your team who have an aptitude for playing this role? You may find your wings can flap faster than you ever imagined. The Cross-Pollinator is an essential part of the ecosystem of innovation. Welcome the role. Encourage it in others. It will help your organization succeed.

CHAPTER 4
The Hurdler

We choose to go to the moon in this decade and do the other things, not because they are easy, but because they are hard, because that goal will serve to organize and measure the best of our energies and skills, because that challenge is one we are willing to accept, one we are unwilling to postpone, and one we intend to win. —JOHN F. KENNEDY, 5/25/61

Hurdlers do more with less. They get a charge out of trying to do something that's never been done before. Lindbergh won a $25,000 prize for his first solo trip across the Atlantic in 1927, flying from New York to Paris on a plane so stripped down it lacked a radio and parachute. Seventy-seven years later, Lindbergh's modern-day equivalent was Burt Rutan and the SpaceShipOne team, who overcame incredible hurdles to launch a civilian aircraft into space and win the $10 million X Prize.

Hurdlers know that you don't always have to tackle a challenge head-on if you can find a way to sidestep it. In the early 1830s, for example, railroad experts playing the role of the Devil's Advocate definitively announced that locomotives could not pull heavy loads up hills, but then imaginative rail entrepreneurs built switchbacks and proved them wrong. And that's a perfect metaphor for what makes Hurdlers successful: Confronted with a road ahead that looks too steep, they approach it from a new angle.

We all know a Hurdler when we see one, the kind of tireless problem-solver who overcomes obstacles so naturally that sometimes it seems as if they weren't even there. Hurdlers can be savvy risk-takers, and are often the most street-smart members of your team. Breaking rules comes naturally, and they know how to cleverly work outside the system. Hurdlers maintain a quiet, positive determination—especially in the face of adversity.

Great Recoveries

At IDEO, our Zero20 group attracts a lot of Hurdlers—or brings out that role in people. They've created plenty of fun and successful toys, from the high-flying Estes rocket that records its altitude to the Fib Finder game that tells whether your opponent is bluffing. It's an amazingly fast-paced business. They pitch hundreds of toy concepts a year, brainstorm nearly every day, and are constantly sketching and prototyping new ideas. The chances of a big hit are low, the budgets tight, the deadlines tighter still. They push the limits so often that it seems as if they are perpetually riding the edge of failure. Like all serious innovators, they have their share of near disasters. But what distinguishes them is they never seem to look down, even when they're walking a tightrope.

How you react to a potential disaster determines your chances of recovery and success. Hurdlers by nature seem to handle challenges the same way great athletes respond to tough competition. Jeff Grant, a key member of our Zero20 team, often demonstrates the Hurdler's knack. One day a few years ago, he flew into New York in anticipation of a critical presentation the next morning with a well-known toy company. Sitting down to a late dinner with a colleague at their hotel in midtown Manhattan, Jeff casually booted up his laptop to review his presentation—and got zilch. The operating system wouldn't start up. They had no backup and no way of accessing the vital material locked within the computer.

There was no time for panic. Toy entrepreneurs don't think that way. But at 11 P.M., the odds of finding and installing a new operating system in time for an early-morning meeting seemed long. Racing through possible solutions in his mind, Jeff glanced out the window and spied a computer store on the street. He jumped into the elevator and ran across the street, but the manager had just closed up shop. From behind the protection of his security gate, the store manager suggested Jeff try Borders bookstore, so off he ran. Borders was still open, but out of the Windows operating system. In the aisles, a tough-looking guy with a linebacker's build overheard Jeff's urgent conversation with the Borders clerk and mentioned that he had a fresh copy of Windows at home. Jeff didn't think twice—even in New York City, where wise people are wary of strangers. He followed the big guy back to his apartment and spent four hours trying to get the operating system up and running. That got him partly there, and at 7 A.M., when the hotel's IT group showed up for work, Jeff enlisted their help in downloading the remaining files. By 7:30, at the last possible moment, the presentation came to life on his laptop, and Jeff sped off in a cab to the meeting. At 8 o'clock sharp, the tired-but-relieved team made its presentation—and sold a toy idea called Real-Action Boxing to Playmate toys, before collapsing in near exhaustion at their hotel.

Jeff's fearless moxie is a great example of the can-do spirit of the Hurdler. Consider how you might celebrate comebacks from the Hurdlers in your group. You'll probably find team members less anxious on high-pressure projects, and more confident of success.

Budget Opportunities

Hurdlers love to turn lemons into lemonade. Give them a constraint, a tight deadline, a small budget, and they're likely to excel. Brendan Boyle used exactly this approach to finding a way around the high cost of bringing his entire team to the industry's annual Toy Fair extravaganza a couple of years ago. Another IDEOer, Paul Bennett—recently transplanted from New York—mentioned that he still owned an industrial-sized loft a couple of blocks from the Toy Fair building, graciously offering the space to Brendan. Originally, Brendan was thinking of attending the show with just one other colleague. Instead, he considered bringing the whole team. They could fly Jet Blue. Stay at Paul's loft. Cook their own food. Brendan ran the numbers and figured that all eight team members could go for less than what it would normally cost for two to fly full-fare coach, stay at a good hotel, and eat out for every meal.

Off they went on their team Toy Fair adventure. They slept in sleeping bags on the floor. They whipped up delicious home-cooked meals. In the evenings, they had furious brainstorming sessions. The camp-like atmosphere transformed the trip into an incredible off-site meeting that stirred the team's creative juices. Staying at a friend's loft apartment instead of a hotel became an advantage, a point of differentiation. The clients they brought back for dinner thought the studio loft was cool. And the eight members jelled as a team. They'd turned the obstacle of a tight budget into a unique opportunity.

How might you turn an obstacle into an opportunity at your organization?

Healthy Innovation

We've seen how Hurdlers make lemons out of lemonade and how obstacles sometimes inspire achievement. The Hurdler's drive plays a major part in significant new innovations and can turn an organization's greatest challenge into its greatest success.

The creation of now-ubiquitous pre-washed salad in a bag—the poster child for healthy convenience foods— is a story of ingenuity in the face of adversity. That whole product category didn't exist back in 1983, the year Myra Goodman (and I) graduated from UC Berkeley. Soon after graduation, she and her husband, both raised on Manhattan's tony upper East Side, took up residence on a small farm in Carmel Valley, near Monterey, and began growing baby lettuce, which they sold primarily to Carmel's popular Rio Grill.

The Goodmans worked the two and a half acres they named Earthbound Farm from sunup to sundown and found they were often too beat to make a salad from scratch at the end of a long day. Sound familiar? They solved their dinner problem by harvesting a big batch of baby lettuce every weekend, then washing and drying it, and putting it in plastic Ziploc bags to be used as part of a healthy meal each weekday evening. They were surprised to discover how long the organic lettuce stayed fresh when properly sealed. "Wouldn't it be great," they mused to each other one day, "if you lived in Manhattan and could get a bag of our fresh salad greens?"

Then adversity struck. The new head chef at the Rio Grill stopped buying their baby lettuce, and the Goodmans unexpectedly lost their only customer. Suddenly they had a field full of very perishable baby lettuce and no one to buy it. But they didn't give up. Instead, the need to drum up new business rekindled the idea of linking their idyllic California farm with the hustle and bustle of their native New York. The Goodmans packaged their washed baby lettuce into little bags, Myra created a nifty label, and they shipped them overnight to Manhattan. Within months, they knew they'd struck something big. They were first to market with a product that is now a $1.5 billion industry. By 2003, Earthbound Farm had grown to become the largest grower and shipper of organic produce in North America—and the most recognized organic produce brand. Today, seven out of ten organic salads sold in U.S. grocery stores are from Earthbound. Sure, Earthbound Farm spurred imitators and worthy competitors, but they still manage to sell $250 million a year themselves, with a product line that exceeds 100 organic salads, fruits, and vegetables. Acreage is up from the original two and a half acres to 24,500. Not a bad recovery from losing your main customer.

The message for food companies is clear: Consumers will eagerly purchase healthy products if they are fresh, tasty, and convenient. But the broader lesson is for companies who feel stuck in mature, low-margin, price-based businesses. If a couple of New Yorkers can reshape a commodity like lettuce into a high-margin product that customers love—while combating obesity in the process—then there must be opportunities in your business, too. Look for a product or service enhancement that you can deliver cheaper and faster. Choose something important to customers, and they will find a way to reward you. Don't despair, even if your original business model is under attack. You may be just one good innovation away from breaking into an entirely new and more profitable industry.

Beyond Adversity to Opportunity

While other corporate executives were lamenting the difficulties of the early-nineties recession, British serial entrepreneur Richard Branson saw a chance to turn the situation to his advantage. He wanted Virgin Airlines to offer seatback video entertainment in coach class to reinforce its image as an innovator in passenger service. Because of the recession, however, Branson reports in *Losing My Virginity* that he couldn't find a banker anywhere that would help him finance the $10 million he needed to retrofit Virgin's aircraft. Many managers would have given up and simply blamed "hard times." Branson thought of another way out of the box. He called Phil Condit, then CEO of the Boeing Company, and asked him, "If I buy ten new aircraft from you, would you be willing to throw in seatback videos?" No one else was ordering aircraft at the time, so Boeing was more than willing to agree to those terms. Branson then called Airbus and made the same proposition. In an economy where Branson couldn't raise $10 million in debt financing for some video systems, he was able to get credit for $4 *billion* worth of new aircraft—and he got his video systems for free. What's more, Branson claimed he never got a better price on aircraft before or since. He not only cleared the financing obstacle, but also got a jump on the competition.

In a similar reversal of fortune, Lexus turned an early problem into a brand-building opportunity. Shortly after Toyota launched their new Lexus brand in the United States, the automaker had a quality issue that was a borderline recall problem. Conventional wisdom says that early failures can permanently damage a new brand while it is still forming first impressions. I can imagine someone advocating "the voice of reason" and suggesting that the company suppress the issue or try to smooth it over quietly. Of course, they did no such thing. Instead,

The spirit of a Hurdler got Virgin its first seat-back video systems—for free.

they contacted every Lexus owner and informed them that the car might have an issue. And what they did next was to flip the situation. To minimize any inconvenience to their new customers, they sent a technician to owners' homes and offices to do a diagnostic and—if necessary—a repair on-site. While they were at it, they cleaned the cars to make them look better than before the visit, something which has become a part of the Lexus formula ever since. As far as I know, the idea of making house calls to every Lexus owner was unprecedented in the history of the automotive industry. Customers sat up and took notice. Instead of damaging the brand, the inspection program gave every Lexus owner the chance to brag about the extraordinary service that came with their new car. Was it an expensive idea to implement? You bet it was. But it turned into a key part of their long-term strategy to build a luxury brand from scratch and raise it to #1 in customer satisfaction. In retrospect, it was a great investment that helped build a lasting reputation.

Outmaneuvering Bureaucracies

One of the myths of innovation is that groundbreaking companies have always been open to new ideas and always give individuals wide latitude to pursue unusual ideas or projects. When we talk about Pixar, Virgin, or Target, there's an assumption that innovative companies have always been innovative. On the contrary, I've found that there's a wide range of organizational acceptance of innovation, even among the most advanced enterprises. When companies do manage to come up with breakaway new products or services, a corporate myth often develops, generally casting the company as the benevolent figure generously assisting employees in their search to come up with the latest and greatest. The reality, of course, is that innovation teams often have to hurdle the barriers set up by well-meaning company management.

Consider the following story of 3M, the maker of the Post-it and thousands of other clever inventions. 3M's tradition of creativity owes a lot to some very persistent and inquisitive individuals. Though you've likely heard all about the origin of the Post-it, you probably don't know the story of how masking tape and Scotch tape came about. It's a story that neatly illustrates how individuals must often hurdle organizational stumbling blocks to create something new.

● HURDLE #1: OVERCOMING THE PRESSURE TO "JUST DO YOUR JOB"

In 1921, Richard Drew, a college dropout playing banjo in a dance-hall band at night and studying engineering from a correspondence course by day, got an entry-level job as a lab technician for 3M. One of Drew's lowly tasks was to take trial batches of the company's Wetordry sandpaper to a nearby St. Paul body shop. On one trip, an auto painter began swearing after ruining a two-tone paint job.

Most lab techs would have considered the painter a grouch and quickly forgotten the incident. But Drew listened. Though two-tone jobs were the rage in the 1920s, the process for achieving the fashionable look was a hodgepodge of newspaper, glue, and butcher paper. No proven method had been developed to mask one color from another.

Drew vowed to the painter that he'd invent a tape to do the job. No matter that 3M was a struggling sandpaper maker with no experience in producing tape, or that Drew was still just a junior lab technician. Drew recognized that 3M already boasted the beginnings of tape: Leave out the abrasive grit when making sandpaper, and you had the necessary adhesive and backing paper.

● HURDLE #2: CIRCUMVENTING COMPANY BUREAUCRACY

With that insight and his drive to solve the auto painter's dilemma, Drew began experimenting with vegetable oils, resins, chicle, linseed, and glue glycerin to create a superior adhesive. When the president finally caught on to what Drew was up to, he ordered him to drop his quest and get back to making better sandpaper. Drew appeared to listen to his superior's request for about a day. As the weeks went by, the president learned that Drew had returned to his passion, but this time he said nothing. Finally, Drew asked for company funds to purchase a papermaking machine to make his tape. The president considered his proposal and then turned him down. But Drew was far from finished. As a researcher, he was authorized to approve purchases of up to $100. So he paid for the machine with a series of $99 purchase orders that slipped under the radar. The result? In 1925, Richard Drew successfully produced the world's first masking tape, a two-inch-wide tan tape with pressure-sensitive adhesive backing. Fittingly, the first customers were Detroit automakers. Far from getting him fired, Drew's insubordination in expensing the prototyping equipment he needed to develop new products came to be seen as a hallmark of the 3M can-do mentality.

● HURDLE #3: SEEING BEYOND YOUR INITIAL FAILURES

A few years later, an insulation firm asked 3M to develop a waterproof seal for refrigerated railroad cars. By chance, DuPont invented cellophane about the same time. Drew immediately wondered whether he could coat the revolutionary material with adhesive. Moving from early prototypes of his see-through tape to a successful product wasn't trivial. The original insulation firm lost interest in waterproof tape. But Drew stubbornly continued tinkering with his dream.

Drew's persistence led to the 1930 invention of Scotch tape, but sales during the first year were a pathetic $33. And the main use that 3M anticipated for the new tape—sealing up packing boxes—appeared to vanish when another company invented a competitive process for heat sealing. The deepening Depression would seem to have been a lousy time to introduce a new product. But paradoxically, the hard times made Scotch tape a staple of the American household. Farmers taped up cracked turkey eggs. Ordinary Americans taped up torn book pages and broken toys. And despite the competition from heat-sealers, many food packers sealed their packages with Scotch tape. Demand was so high during World War II that shortages were common.

If Drew hadn't consistently bent or broken the official rules, none of this would have happened. He seemed instinctively drawn to novel products, fighting his own company's institutional reluctance to risk-taking. He was a true Hurdler in the broadest sense: one of those rare innovators responsible for not one but two major successes, because he heeded a need from real customers and pressed forward, despite the concerns of management or an uncertain market.

Not surprisingly, Richard Drew ultimately became a 3M legend, emblematic of the company's well-earned reputation for innovation. In doing so, he proved that individual Hurdlers play a huge role in changing an organization's view of innovation. Indeed, the true strength of a company is often the force of ambition and imagination of its most creative—and occasionally most stubborn—individuals. And as a serial innovator, Drew proved that lightning can strike twice within a culture of innovation.

Real Hurdlers

Real-life Hurdlers are a wonderful source of inspiration, because the metaphor of someone hurdling over a barrier is so visual and so apt.

Perhaps you've seen some hurdling feats in track and field during the Olympics. Truly, they have to be seen to be believed. Great hurdlers run nearly as fast as they would without the barriers. Consider the 400-meter hurdles, one of track and field's most demanding events. The race requires speed, balance, perfect choreography, endurance, and guts. When Edwin Moses began his stunning decade-long dominance of the event in the mid-1970s, virtually every other hurdler in the world took fourteen steps between each of the ten three-foot hurdles. Moses took thirteen, going against conventional wisdom. He also brought scientific discipline to his training routines. His results were fantastic. Between 1977 and 1987, Moses won a phenomenal 122 straight races. His 1983 world record of 47.02 remains the second-fastest time in history—more than two decades later. In a 400-meter race *without* hurdles, Moses came in around 45 seconds. In other words, leaping ten high barriers around a quarter-mile track added scarcely more than two seconds to his time.

Great Hurdlers hardly let obstacles slow them down, much less stop them. Which goes to show that a hurdle is only as high as you make it out to be.

Olympic gold medalist Edwin Moses barely let the hurdles slow him down.

Hurdling with a Blog

Diego Rodriguez is one of IDEO's boomerang staffers. He left IDEO to get a Harvard MBA, then worked at Intuit for a while as a brand manager for QuickBooks Online. While part of the marketing team at Intuit, Diego had an idea that he thought would help draw attention—and ultimately more customers—to his online service. Diego understood the power of blogging and wanted to apply it to his product line. Most of the marketing group didn't see much potential in blogging (just as, a generation before, many people pooh-poohed the value of the Internet) and warned Diego it was a waste of time. "It's not on our 'critical few' list," he was told, which is the Six Sigma–speak equivalent of "don't bother."

The blog cost a whopping $13 a month to host, and that was more of a barrier than you'd expect—for two contradictory reasons. The first reason, almost unbelievable in a company of Intuit's size and success, is that it was beyond what Diego could get authorized going through normal channels. "It was near the end of the fourth quarter, and there was just no budget *anywhere,*" reports Diego. The second reason, ironically, was the opposite: "I didn't think I could convince anybody that something really important to our marketing efforts would cost only $13 a month." One Friday at lunch, Diego got a friendly executive to put the charge on his personal credit card, although virtually no one else in the company knew what he was up to. Diego spent the weekend writing a few dozen entries for his blog. Meanwhile, he called a freelance graphic designer to ask if she could come up with some simple graphics by Sunday night. A few days later, he was up and running with a blog that discussed his favorite QuickBooks features and gave expert tips on using them. Diego somehow got a link to his blog onto a page of Intuit's main Web site. In the first couple of days, he was gathering new visitors at the rate of about one a minute. He sensed he was on a path to something great. Or about to get fired.

Intuit's corporate marketing department put out an advisory that the company would not be getting into blogging, apparently oblivious to the fact that they already had an active one. The "blogosphere" was a pretty close knit group back in 2004, so when blogging luminary Robert Scoble noticed that Intuit (or at least one renegade there) had a blog going, he praised the company for being an innovation leader and created a link from his site to Diego's. From that moment, Diego's little blog developed a life of its own.

More and more customers started finding it and leaving positive feedback, writing things like "This is so cool." Like many Web-savvy companies, Intuit sometimes pays to have their site listed among the paid search results on Google. But suddenly their Google hits went higher, as more bloggers and other sites linked to Diego's QuickBooks blog. By the time Diego's management team tumbled to the fact that they had a blog at all, it was already a phenomenon. Everyone loves a success, of course. Diego's managers came around to thinking it was a darned good idea too, even if they quietly worried about his end run. But the point is, Diego is a Hurdler. If Diego had taken no for an answer the first time, if he hadn't had the heart of a Hurdler, it would never have happened. And that is the spark that a Hurdler brings to their team.

Hurdling Beneath the Radar

Here's a Hurdler story from Apple Computer that would be dismissed as an implausible Hollywood screenplay if it weren't actually true. Ron Avitzur now heads his own company, but back in 1993 he was a software contractor at Apple. When the big project he was working on got canceled, Avitzur was so passionate about his part—a graphing calculator to help students visualize and solve geometric math problems—that he kept working on it anyhow. When he realized that his contractor's badge still got him in the front door at Apple, Avitzur set up shop in an empty office and became a one-man skunk works project. Later, his friend and fellow Apple contractor Greg Robbins also lost a project, and the graphing calculator team doubled in size. They continued working seven days a week, twelve hours a day, trying to make their dream a reality. No

No budget. No official approval. In fact, no authorization at all, but in true Hurdler spirit they kept moving forward.

budget. No official approval. In fact, no authorization at all, but in true Hurdler spirit they kept moving forward. Of course, eventually they were discovered. One day a facilities person came to inspect their "empty" office and discovered Avitzur and Robbins occupying it. "Who are you guys?" she asked, calling security, canceling their badge numbers, and asking them to leave the building. Which they did, of course, but with the Hurdler spirit, they were back the next day, getting past front-door security by "tailgating" in behind real employees.

They continued sneaking in for weeks, eventually winning the support of an underground network of Apple engineers who admired what they were doing and wanted to help. They got access to rare prototype machines so that they could test their software out on Apple's not-yet-released PowerPCs. They got graphic-design help, usability testing, QA, documentation—all the specialties that go into making the production version of a solid new software tool.

They knew they couldn't stay underground forever if the graphing calculator was going to actually ship, so eventually they did a demonstration for a select group of Apple managers. With the help of the grassroots team, Avitzur won over the group of managers that day. Amazingly, he still didn't have official status. So he and Robbins continued sneaking into the building a few more weeks until an engineer took pity on them and figured out a way to navigate through the obstacles and get them "vendor" badges.

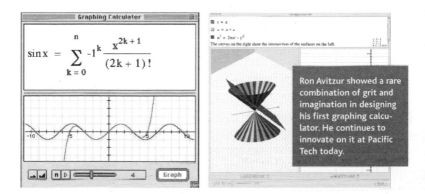

Ron Avitzur showed a rare combination of grit and imagination in designing his first graphing calculator. He continues to innovate on it at Pacific Tech today.

Eventually, the race was run, the hurdles hurdled, and the graphing calculator shipped on 20 million Macintosh machines. Avitzur is the CEO of Pacific Technology, a company that continues to create better new versions of the graphing calculator software, which he now licenses to Apple. He doesn't have to sneak in the door anymore.

The Power of a Constructive "No"

When I think of a Hurdler, I think of someone with extraordinary resilience, someone who doesn't take no for an answer. But sometimes even a Hurdler can use the power of "no." There are times when you have to reject an initial strategy in order to find the right path for a new idea. A few years ago, for instance, Kevin McCurdy, a successful young Internet entrepreneur, came to us with a compelling idea. He walked in with a Sony Vaio notebook computer and showed us a magazine he'd scanned into the computer's memory. He had a simple if ambitious goal. He wanted to own the emerging world of digital magazines.

McCurdy knew IDEO had designed the Softbook, one of the world's first electronic books. Since we'd done it for books, why not for magazines? As a wealthy entrepreneur, McCurdy could fund the development himself. There would be no bureaucracy to cut through.

So we began prototyping. There were technical issues. The size and shape of screens, the length of battery life, not to mention the nuances of how people read and engage with magazines. As our team pushed further into the project, it seized upon strengths in the digital medium. Publishers wouldn't have to pay for distribution or printing, and the advertising could boast intelligence and interaction. Today's average of eight magazine subscriptions per household would offer a large target market.

But as we began prototyping a dedicated magazine reader, we realized we should say no to hardware. We told McCurdy that he'd have a tough row to hoe in the hardware business. There were numerous reasons why introducing a magazine-dedicated device would likely

fail, from the high cost of manufacturing to the difficulty of generating sales
with a stand-alone device. Long experience has taught us that the hurdle
for novel hardware is high. We
worked on a number of pen-based
computing devices that broke more
ground than sales records before we

McCurdy, a veteran
entrepreneur, didn't
consider the possibility
of failure. But he was
uncommonly flexible
about how he would
succeed.

teamed up with Palm Computing on the hit Palm V. Sure, we could
make a beautiful magazine reader, but like the Softbook, it would probably be ahead of its time.

McCurdy, a veteran entrepreneur, didn't consider the possibility of
failure. But he was uncommonly flexible about how he would succeed.
He was strong enough to see the early hardware mock-ups for what
they were—informative prototypes—and wise enough to understand
that in some way they moved us toward the real opportunity. None of
us gave up on his vision. Together, we designed a software experience

Zinio sidestepped
obstacles on its way to
delivering more than 26
million digital magazines

for online magazine subscriptions. That wasn't easy. The online versions had to have the look and feel of their print twins.

By the spring of 2002, his company, Zinio, had launched a Web site and service. Today the Zinio Reader is considered one of the best in the industry. Zinio has ample competition, but the company is already profitable and has gained considerable momentum—they've delivered more than 26 million digital magazine issues in more than 200 different titles, from *BusinessWeek* to *Cosmopolitan*. The advantages of online magazine reading are clear—easier searching and archiving, not to mention reading on the move: Subscribers in more than a hundred countries read Zinio editions.

McCurdy deserves most of the credit. He had the vision and the drive. We had the gumption to tell him that a grand hardware design wasn't necessary. He had the strength and flexibility to adjust his course.

What's the lesson? Sometimes to say yes, you've got to first say no.

No Cash? Print Your Own.

Clever innovation in overcoming hurdles is not the exclusive domain of start-up companies or boutique brands. Take, for example, Cargill, one of the largest food companies in the world, with more than 100,000 employees spread across the globe. Faced with a financial crisis in Zimbabwe a couple of years ago, Cargill encountered a hurdle you probably haven't ever had to deal with: Not enough currency existed in the local economy for the company to pay cotton farmers for their crops. After exhausting all ordinary means of finding more currency, Cargill took the remarkable step of *creating its own*. Technically, the beautifully printed Cargill notes were fixed-denomination bearer checks. But they became de facto legal tender in a country where the national currency had been ravaged by inflation. Within a few days of being printed, the notes—with a value of roughly $1 each—were being accepted by wholesalers and retailers for all kinds of cash transactions. Although the notes were obviously not official government currency, the Cargill name was enough to make them as useful as money. And bearer checks helped farmers get fairly paid for their harvest.

Macro Hurdles and Micro Mills

Anyone who has actually lived through a wave of breakthrough innovation can tell you it is anything but inevitable. The timing and outcome of noble experiments is never a sure thing—which is why they're considered experimental. So look around and ask yourself which ventures currently under way represent a future that's arrived slightly ahead of schedule. One such effort that looks like a good bet to me—but has been surprisingly slow to catch on so far—is the advent of micro mills for making aluminum soft-drink cans. As you may know, mini mills in the steel industry have innovated their way to industry success, as low-cost producers like Nucor systematically stole market share (and market capitalization) from giant integrated steelmakers like Bethlehem Steel. Now a revolutionary new technology threatens to do the same thing in the aluminum industry. Gavin Wyatt-Mair and Don Harrington at Kaiser Aluminum have pioneered a process that collapses the industry's fourteen-step, two-week production cycle down to a half-day micro-mill process. Micro mills cost half as much as traditional mills to build, boast a lower total cost per ton of output, and are small enough to locate next door to a bottling plant.

What's the catch? Kaiser built its first prototype plant in 1997, and the aluminum industry has yet to embrace the new technology. Still, I wouldn't bet against it. Some breakthroughs are a long time in the making, but when they hit shore, they're often well worth the wait.

I remember, for instance, first seeing digital light-processing technology (DLP)—then known as a digital mirror device—on a visit to Texas Instruments back in 1991. Originally developed by TI's Larry Hornbeck in 1977, the technology matured at what seemed a glacially slow rate. It was nearly twenty years—not till 1996—before the first commercial units shipped. But today that slow start is all but forgotten. DLP has become the gold standard in projecting computer images. Texas Instruments has more than 100 patents on the technology and has shipped more than 5 million DLP-powered units.

The lesson? Sometimes it takes time for the biggest new ideas to upend the status quo. When an industry has a large investment in infrastructure—outdated or not—you've often got to stay the course for many years before your new innovation can take hold.

Silver Linings

Part of the Hurdler's role is trying to find the silver lining in every cloud. Setbacks aren't problems, they're opportunities. Here's an example from Yoo-hoo, the New York–based maker of retro-chic chocolate soft drinks. One day a Yoo-hoo beverage van was stolen on the streets of New York. What possible good could come out of such a theft? Well, the folks at Yoo-hoo, known for their whimsy, posted a mock all-points bulletin on their Web site. They put up a hefty reward—two years' worth of chocolate Yoo-hoo, encouraged visitors to their site to download "Missing" posters, and generally turned a routine misadventure into a marketing opportunity. The campaign might have lasted longer if the New York police hadn't located the abandoned van a few days later.

I think that more companies could learn from Yoo-hoo's example. When little things go wrong—mix-ups and minor mistakes—most companies would be better off treating them with a sense of humor. Most publicity is better than no publicity at all. So the next time something goes awry, consider how you might turn it to your advantage.

Perseverance Pays

The essence of a Hurdler is perseverance. Recently, I heard a vintage never-say-die story firsthand from former Polaroid staffer Burt Swersey, who worked there in the early days. Back when Polaroid's instant photography was still a hot new technology, one of the more colorful characters there was a sixty-four-year-old engineer by the name of Sid Whittier. Sid had become a millionaire several times over from his Polaroid stock, but the eccentric senior citizen still showed up for work every day, usually dressed in the same tattered sport coat.

Sid did some early design work on Polaroid's new Model 150 camera. He started in on developing an innovative and very precise shutter mechanism—an idea he thought had great merit—but it was passed over for an alternative design done by one of the young engineers down the hall.

Months later, after Polaroid had conducted extensive further development on the competing design, it proved impractical to manufacture, leaving the company faced with a painful project restart—along with costly production delays. Polaroid executives were tearing their hair out, trying to imagine a way out of their predicament. At that climactic moment, Sid Whittier reached into his desk drawer and pulled out a completed design of his original shutter concept. Sid believed so much in his idea that this semiretired millionaire engineer had continued working nights and weekends to refine and perfect it, even after it had been officially rejected. Against all odds, Sid's dogged persistence saved the day for Polaroid, and they took his design to market.

Experts on the Status Quo

The lesson I draw from the Polaroid tale is that sometimes it pays to maintain optimism when being pulled down by those who should "know better." I'd be the last person to suggest that you completely disregard the experts. But, by nature, experts are often the guardians of conventional wisdom. They have deep knowledge of what has worked in the past, and that knowledge can be extremely valuable. But sometimes a new idea or method, or simply a fresh environment, can make the old views suddenly look outdated.

Both my brother David and I—four years apart—had similar experiences overcoming well-informed naysayers during our senior year of high school. We both applied to challenging colleges, hopeful but uncertain that we would succeed. Each of us had an "expert"—someone we trusted—tell us to lower our sights. David's high school guidance counselor told him that Carnegie Mellon University was "too hard and too far away." Four years later, mine told me that Oberlin College would "eat a small-town boy alive." "Play it safe," they were implicitly saying. "Stay close to home, go to the nearest school, don't get your hopes up. Don't aim too high." Experts like that are usually at least partly right. God knows, they have a lot more data than the average person. They can quote chapter and verse about "the way things are done." Carnegie Mellon *was* incredibly difficult, and David struggled with the math and physics that hit hard from the first day of school. Oberlin *was* an eye-opener for someone who had never even eaten a bagel or tasted Chinese food, let alone read Ivan Illich or Henry David Thoreau. But shame on them for telling us we couldn't do it, for trying to reel in our youthful enthusiasm. And God forbid we had listened to them. David went on to become a full professor with an endowed chair at Stanford—not to mention creating IDEO from scratch. And Oberlin gave me a thirst for learning and a love of writing that have helped me enormously throughout my life.

So nurture the Hurdlers in your organization. Cherish them. Hurdlers can seem at times a bit stubborn. But that can be a good thing. The Hurdler listens to experts but doesn't let them have the final word when it comes to his or her own thinking, career, and life. Ignore the experts and sometimes the walls in front of you will turn out to have doors.

Then you can find your own path.

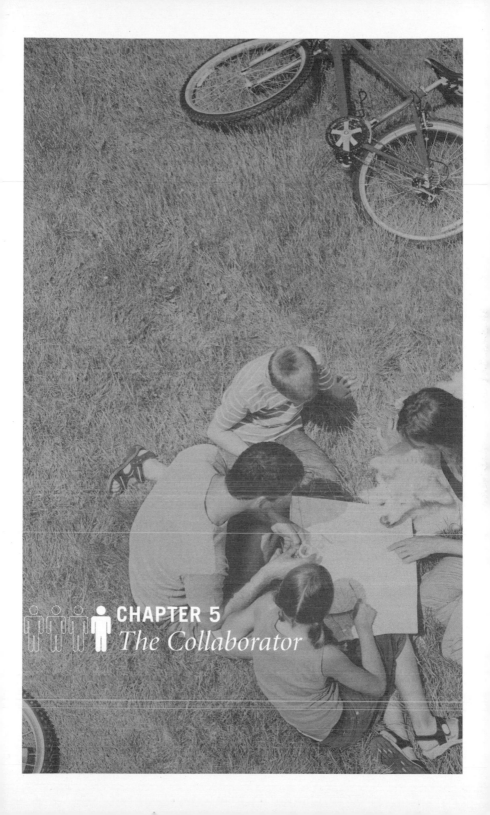

CHAPTER 5
The Collaborator

In the long history of humankind (and animal kind, too) those who learned to collaborate and improvise most effectively have prevailed. —CHARLES DARWIN

Thomas Edison went down in history as America's greatest inventor partly because he was a master Collaborator, championing and cheerleading a talented and coordinated team that churned out a tremendously wide-ranging series of inventions and innovations. Similarly, when England needed to crack Germany's Enigma crypto machine in World War II, it gathered together a diverse group of scientists, mathematicians, and engineers at Bletchley Park, creating a brain trust that solved the single greatest intelligence challenge of the war.

What is a Collaborator? We all know them. Collaborators stir up the pot. They bring people together to get things done. They're proactive cross-trainers, willing and able to leap organizational boundaries to coax us out of our silos to work together in multidisciplinary efforts. They dream up multilateral task forces and make them work. On teams, they often lead from the middle, using their diplomatic skills to hold the group together when it threatens to splinter or disband. When energy or enthusiasm flags, there is no better cheerleader.

IDEO itself was formed through an unusual collaboration. When I joined the firm, it was called David Kelley Design, a small engineering-oriented firm founded by my brother David. We really did want to change the world—at least a little—and my brother decided around 1990 that we'd have a much better chance of achieving our dreams if we were more multidisciplinary. So in 1991, he engineered a three-way collaboration between Bill Moggridge's firm, ID Two (in San Francisco and London), and Mike Nuttall's Matrix Product Design. Collectively, our firms had collaborated on more than sixty projects. But bringing these complementary (and occasionally competitive) firms together in a three-way merger was still a pretty radical idea. Yet the new collaboration went smoothly from the start. The spirit of cooperation, built up through the dozens of projects we'd worked on together, paved the way for a painless transition.

In the last decade and a half, perhaps the distinguishing characteristic of IDEO has been our ability and eagerness to collaborate with an increasingly broad mix of companies and organizations. We've helped hospitals improve the patient experience; fine-tuned the design of planes, trains, and automobiles; helped leading universities innovate on the learning process; and even sparked new thinking at the Internal Revenue Service. As one of our favorite clients kindly put it, "What IDEO may do best is collaborate."

But there was a time when IDEO might have considered that comment a slight rather than a compliment. When we were primarily a product-design firm, we sometimes saw our role as coming up with the right solutions and handing them over to the client. Today, we more often see ourselves as working alongside a client group, influencing their culture, altering their patterns of innovation, and leaving them with new tools to continue the forward momentum. I know from a former career in management consulting that the "right" answer is worthless if it is rejected by organizational antibodies in the client company before implementation even begins. When it becomes difficult to distinguish our contribution from that of our clients, we know we're on the right track. In the last few years, there has been a new benchmark for IDEO project teams: The ultimate proof of success is when our counterparts at the client company get promoted.

So who is a Collaborator? What role do they fill? That rare person who truly values the team over the individual, the project accomplishments beyond individual achievements. The person willing to set their own work aside temporarily to help you make a tight deadline. The person you can count on to jump in when and where they are needed most.

Collaborators can be a company's best defense against internal skeptics. A top-down approach often galvanizes the resistance of a skeptical staff member. But a savvy Collaborator can perform a subtle form of corporate jujitsu, ultimately turning the strength of any initial opposition into a positive force. We've watched this happen on dozens of projects. One or two members of a joint-project team announce that the project is a waste of time. If a Collaborator can win over those skeptics, they can become our best friends. With the zeal of converts, they go on to champion the process as much as the project. It happens more often than you would think, and when it does it is supremely rewarding.

> But a savvy Collaborator can perform a subtle form of corporate jujitsu, ultimately turning the strength of any initial opposition into a positive force.

Collaborators know that the race is won in the baton pass. They excel in the handoffs between departments and team members. Sometimes you can spot a Collaborator right away. That was true of Maya Powch, one of our youngest employees. Midway through her job interview, we realized that there was a project we could use her help on starting that same afternoon. We said we'd get to work on the offer letter if she'd begin work immediately. Her intense, two-month project was aimed at informing new-product strategies through studying insights into eating and snacking patterns in Latin America. As a new employee, Maya was not surprised that her responsibilities included entry-level tasks like gathering prototyping supplies and making simple mock-ups. But she found she was also an integral part of the design team. When they needed an extra human factors person to perform observations in Mexico, Maya volunteered. And when the project manager left the team to form his own company, Maya stepped in and gave the critical final presentation to the client in Mexico City. She even brushed up on her high school Spanish to make sure she'd pronounce everyone's name correctly. It was a great success. Collaborators speak the team's language and come through in the clutch.

Unlikely Partners

Especially if you live in Silicon Valley, giant food companies and grocery store chains aren't usually the first groups that come to mind when you think of innovative business collaborations. But we've seen them come up with breakthrough process innovations—when they come to the table in a spirit of collaboration.

Not long ago, Kraft Foods and Safeway had a pretty typical vendor-retailer relationship. Which is to say that it left plenty of room for improvement. But Ron Volpe, supply-chain director for Kraft, believed he could turn things around if he could tap into the collective knowledge and energy of both companies.

His challenge was similar to that facing countless supply-chain managers around the world. Volpe's day was consumed with small skirmishes, mustering on-the-spot responses to short-term tactical problems: a truck that didn't arrive on time, a mixed-up shipment, or a pricing dispute. Just getting through the daily barrage of tense phone calls and hundreds of e-mails was a continuous struggle. Volpe had little time or resources to develop new long-term strategies. So what did he do? He came up with an innovation opportunity.

Volpe contacted Linda Norgren, his counterpart at Safeway, and asked if she was open to the idea of a joint innovation project. As he put it, why not take down the walls, drop the barriers, start together with a blank sheet of paper, and see how we might improve. Norgren was interested, but even so, the project almost didn't happen. Several of Volpe's own salespeople wouldn't commit to a time. A get-together had to be postponed *twice* because so many people from both companies backed out.

After months of juggling schedules, the Kraft and Safeway teams finally got into a room together at IDEO Palo Alto, with my colleagues Peter Coughlan and Ilya Prokopoff from our Transformation Practice. The collaboration that day was designed to find ways Kraft might better link its Vendor Management Inventory (VMI) system with Safeway. Essentially, VMI allows special suppliers like Kraft to manage the inventory and ordering of its products sent to Safeway's warehouse distribution centers. When it works smoothly, there's a perfect balance—neither too much nor too little product. As the collaboration got under way, however, the two companies decided to broaden the discussion. They saw a potential for doing more than improving VMI.

Even before new ideas started emerging, both teams got to know one another—no small feat in this age of high-tech communications. Volpe could name no more than half the Safeway team members, and suspected that most of the retailers knew even fewer of his staff. They brainstormed and divided up into groups investigating different topics. They got a little wild. More than a few high-spirited individuals scribbled out Post-its that made light of their respective bureaucracies. They drew Oreo cookies as organizational pie charts and fired our spongy Finger Blasters at each other, blowing off steam. Before the day was over, they had generated some 133 ideas. They assigned joint Kraft/Safeway task forces to five major topics. They resolved to establish aligned goals and launched a concerted effort to collaborate on strategy, promotions, logistics, and real-time data.

The concept was for the teams to quickly develop a number of prototypes. That might have been it—a promising opening session, then a slow fade-out as both teams returned to their companies and their traditional ways of doing things. But Volpe did something simple and powerful. He got permission to gather up everybody's e-mail addresses. Credit Safeway for that openness. Many companies wouldn't have allowed such unmonitored cross-communication. Long after the excitement of the kickoff had passed, a series of reminder e-mails helped Volpe to keep pressing the task force teams forward on the main topics, asking for weekly updates. Some teams faltered, while others kept up their momentum. Meanwhile, IDEO led "experience audits" at Kraft and Safeway to get a candid view of the operating environment. One critical proposal emerged for streamlining the supply through what's known in the industry as "cross-docking." Whereas most items were shipped to a central Safeway warehouse, placed into storage there, and then shipped out to stores as needed, Kraft and Safeway developed a strategy to transfer Kraft products straight across the loading dock into a Safeway truck. Cross-docking can reduce handling costs, tighten up the supply chain, and increase inventory turns. The streamlined approach saved money for both companies in labor and carrying costs.

As part of making the new distribution tactic work, Kraft changed the way they palletized their products. The new pallets required no storing or shelving and were immediately ready for sale. Kraft and Safeway chose Capri Sun as the first product on which to use their new process, and within weeks the drink started flying off the pallets. They achieved 100 percent cross-docking, and revenues soared, increasing by 167 percent. It was far and away the best Capri Sun promotion at Safeway ever, and both companies were extremely pleased with the results. Wondering if it was a "one-hit wonder," they repeated the process with a promotion for Kraft's flagship Macaroni and Cheese product, with stellar results again. Meanwhile, Kraft was on its way to becoming a very popular Safeway vendor.

By putting their heads together, the two companies essentially created a ready-to-go display system while reducing the need for handling. The super sales led to good vibes. Suddenly, both companies were enthusiastic about collaborating. Success led to a mutual desire to measure just how well they were doing. They wanted better metrics to track the system. At the end of the first year, they hatched an idea to create a jointly developed scorecard for the retailer to rate Kraft's performance. Volpe's group helped bang out the first prototype, a quick-and-dirty daily e-mail. Within a couple of months, Kraft and Safeway agreed on sixteen discrete measurements to plug into a spreadsheet. Kraft led the initial work and Safeway refined and formalized the tool.

The resulting Supply Chain Scorecard gave Safeway a powerful way to measure Kraft's performance. Safeway was so impressed by the system that it expanded the tool to rate its top twenty-five vendors. Kraft was rated as a "Gold" vendor, but more important, the company had proved it was a true partner. Safeway and Kraft celebrated the joint efforts in internal publications, and the Grocery Manufacturers of America presented Kraft with its CPG Award, which honors product innovation and industry collaboration. Since then, the companies have gone on to simplify purchase orders and invoices, as well as standardizing the terminology they use in reports. Today, they are working together on a variety of fronts. The goal of one major initiative is to keep Kraft products from going completely out of stock, an area where slight improvements reap multimillion-dollar results.

To me, this story demonstrates how radical collaboration can dissolve traditional barriers between vendors and their business customers. It's an approach that can be fruitful for any number of business relationships, even those that appear at times to be adversarial. Taking down some of the walls separating inter-company teams can make a world of difference.

Unfocus Groups

Focus groups. The phrase is almost synonymous with marketing in the second half of the twentieth century. Focus groups bring together typical customers to give companies verbal feedback. They can be valuable at the validation stage, but we don't believe in focus groups if you're looking to inspire breakthrough innovation. We don't think you learn much from the "usual suspects" when you're trying to create something new-to-the-world.

We do *Unfocus* Groups. We invite extreme people passionate about the products or services we're trying to develop. Unfocus Groups offer inspiration on innovative design themes and concepts. They provide human grounding for designers and project leaders. They also demonstrate in a physical, tangible way what truly excites and drives people.

We've held Unfocus Groups on projects ranging from shoes to banking, consumer electronics, and cars. Dorinda Von Stroheim is our casting director for these sessions. She's got an uncanny knack for finding fascinating, animated, verbal, and offbeat participants. The cast for a recent Unfocus Group about shoes was remarkable for its eccentricity: It included a lounge singer wearing a fedora with a thing for shiny loafers, a mom with a favorite pair of racy thigh-high leather boots, and a man who literally walks on fire and loves his sandals. How do we know what shoes they wear? Dorinda asked each of the eight participants to bring in a couple of pairs of shoes. Suzanne Gibbs, who led this and many other IDEO Unfocus Groups, calls this "homework," giving the attendees something to do to prepare them for the session.

Suzanne began the three-hour event around 5:30 P.M. at our San

> We don't think you learn much from the "usual suspects" when you're trying to create something new-to-the-world.

Francisco office. After some brief initial creativity exercises, the participants showed the shoes they'd brought and talked about their feelings for footwear. An artist who doubled as a limo driver talked about how she dressed professionally and expressed her personality in her expensive and carefully maintained black heels—a side of her that passengers saw only when she got out of the limo. The firewalker spoke of how he loved open-toed sandals. Suzanne went on to introduce some design themes IDEO had developed through its work on the project—Relaxation, Secrets, and Invisibility. She split the room into small groups and asked them to prototype a shoe that represented one of those themes.

This may be the biggest difference between Unfocus Groups and the traditional approach. We ask extreme and exceptional people to bring their passions and interest to prototyping. In this project, we were exploring health and wellness. Our participants made shoes that accentuated the healthy aspect of the inside of the shoe—the heel cup, the toe grip, and the arch support. Others made shoes that represented whimsy and lightness, or that explored secrecy and intimacy. One team built a prototype black pump with a secret drawer in the heel. The prototypes gave us a clearer sense of some of the qualities people desire in shoes. Together, these prototypes and the group's insights made for good grist for our designer. New Zealand–born IDEOer Joanne Oliver created some concepts for a sandal that elegantly accentuated form and fuction. Our client went on to design innovative new footwear that expanded the fashion sense of a brand that had originally emphasized only practicality.

Unfocus Groups flesh out design themes and points of inspiration. They mix elements of observation, prototyping, and brainstorming. They're also a mini performance. On another project, for example, instead of asking the group to describe how they share food in social settings, we had them prototype social food packaging. While exploring banking, we had individuals bring in something that represented their approach to finances, which ranged from impeccably organized binders to haphazard shoe boxes crammed with receipts. For an automaker interested in cars for youth markets, we invited a colorful group to show off their wheels—everything from souped-up street machines to exotic foreign models and a giant SUV (which the owner used to carry around his paraglider). They parked one night in our waterfront warehouse on San Francisco Bay, where we turned on the floodlights to let them open their hoods and rev their engines. On a project to develop dog-grooming projects, we invited pet owners to bring their canines—and quite honestly saw it all: the first office dog fight, dogs urinating on our floors, and a couple sitting on our "look-out" couch, showing off their grooming products while a large mutt trod back and forth on their laps.

Quirky, fun, and often surprising, Unfocus Groups give companies a chance to see real people interact and experiment with products and things they care about. Suzanne sees it as a more authentic way to engage with consumers. Give it a try the next time you want to check out some new design objectives or get a tangible sense of what consumers might care about. Remember to cast your group with a wide range of characters—people with passion and highly unique interests. You'll likely find a deeper sense of purpose about the individuals who make up your customer or client base. Unfocus groups put faces and emotions on the deeper underpinnings of products and services. They add another human level to the process of innovation.

Cross-Training

Thousands of collaborations with a broad range of companies have given us insights about the inner workings of corporate teams. Companies typically have departments that excel in their respective competencies, such as marketing, finance, engineering, and manufacturing. Many large organizations are structured around discrete and separate functions or geographically separate offices. What we've found is that trying to fit a promising innovation project into these existing management silos can be like trying to slice a 360-degree panorama into a series of small frames. You never quite capture the full perspective. At some very good companies, the best they can hope for is a series of handoffs that don't wreck the momentum. If you're riding the wave of previous success, this can keep you afloat for a while. But management built on dogma can run into trouble when the business climate changes. You can't mechanistically build responsiveness into a company. Or innovation. For example, companies in the health care industry need to constantly change and adapt in response to pressures caused by new regulations and emerging competitors. Ditto for industries ranging from fashion retail to continuing-education services. For organizations split into silos, their siloed thought patterns pose a stiff barrier to the breakthrough innovation the market demands.

To illustrate how this often happens, let me tell you about a day I spent a few years ago with a large company in the stationery business. On a tour of their headquarters building, we peeked into a room dedicated entirely to cataloging their competitor's products. Arrayed on the floor-to-ceiling shelves were hundreds, if not thousands, of cards and other paper items made by competitors. I am a big fan of such collections. They remind you of the diversity of innovative solutions happening *outside* your organization, making it harder to get lulled into thinking you've cornered the market on good ideas. In our San

The irreverently named "Packaging Hall of Fame" reminds us that clever ideas can come from anywhere in the world.

Francisco office, for example, we have our Packaging Hall of Fame, a wall full of hundreds of quirky and cool and thought-provoking shampoo bottles, soda cans, snack containers, and the like, collected from the far corners of the earth.

But as I looked at the mass of competitive products from this stationery company, I told the curator of the collection that I noticed one significant omission. There was nothing from Japan. I've been to Tokyo more than two dozen times and never failed to visit a couple of stationery stores, because I find their products so creative and well crafted. "That's a great idea," piped up the curator. "How would I do that?"

"Well, the way I would do it," I said to her, "would be to call one of my friends in Tokyo and offer them a few hundred dollars of budget to buy the latest products from the remarkable Tokyu Hands store in the Shibuya district or Itoya's Ginza stationery shop. They'd FedEx the stuff to me and I'd have a whole new collection within a few days." The curator seemed to like that idea, and implied that she might ask me to do that for her. But then, as we were walking away, my host and tour guide for the day whispered conspiratorially, "You know, Tom, we have a big division in Japan."

I was stunned. The company had several hundred workers in Japan, but the curator at headquarters didn't quite know how to call or whom to ask. It's an example of how silos can hold you back. They've got plenty of resources in Japan but can't seem to tap into them.

Of course, we've often played the role of connector, helping to bridge one part of a client company to another. In a perfect world, companies wouldn't need such help. But I believe you've got to do whatever works. Sometimes companies need a third party to help facilitate interdepartmental collaboration.

How do we combat this traditional silo view when we collaborate? We create a cross-functional team. It's more than simply inviting members of the various departments that see a concept through to a new product or service. Cross-functional teams can orchestrate jam sessions between departments that often don't talk to one another, let alone play music together. I'd be the first to admit that this can be tough going. Egos are involved as well as turf. But I think it's one of the central elements of a successful corporate collaboration. At IDEO, we're designers, of course, but we also see ourselves as advisors and tour guides on this journey. We're in the room and around the table. We work to help make it a unified team and not just a collection of representatives from separate teams.

Recently, there's been a lot of talk about the importance of multidisciplinary teams. But simply including someone from every team isn't enough. You need some glue. You need a Collaborator. Someone who coaxes people out of their silos and patterns of thinking into the uncharted territories that hold promise. That's not an easy objective, but it is a worthy goal. And the effort of the Collaborator to make it possible will be time well spent.

Every team needs at least one Collaborator. Preferably, a lot of them.

Collaboration Through Cohabitation

Why do collaborations matter? Where's the evidence that it makes sense to partner, to team up, to work jointly toward a goal? Over the years, we've worked with a lot of companies on highly collaborative projects that were successful. We've had Japanese designers from Matsushita spend weeks in our London studios. We've had an IDEO team work inside BMW's Munich research center for a year. We've helped companies like Kodak and Procter & Gamble set up innovation centers to spark new ideas. One of our less-celebrated collaborations, however, may be the most illustrative of the power of teamwork.

More than ten years ago, Samsung came to us with a bold plan. Then widely considered a second-tier consumer electronics company, they had efficient manufacturing capabilities but were not particularly well known for good design. On their own initiative, they sent a rotating group of designers from Korea to work and essentially live with our designers in California for almost three years. Together, we designed twenty-seven new products—from computers to televisions.

Samsung built a compelling brand with the help of international alliances.

Samsung built up its design capability on both sides of the Pacific. Before long, their striking new designs were making the cover of *Business Week*. Sales were superb. Once undistinguished among Korea's giant *chaebols,* Samsung is now the indisputable leader, ahead of companies like Hyundai and Lucky Goldstar. Interbrand's annual survey of top corporate brands recently rated Samsung as one of the fastest-rising brands in the world.

Of course, IDEO had only a small part in the transformation that led to all that success. Samsung has continued to invest heavily in R&D, fine-tune its efficient manufacturing, and improve its marketing efforts. Still, I think it's fair to say that our joint collaboration played a catalyst role in solidifying Samsung's commitment to the power of design. And like the best collaborations, it's not over. As I write this, a Samsung designer is living with us once again.

Triathlon Is the New Golf

Golf has always been considered the ultimate business-oriented game, with plenty of opportunity to talk shop on the course and, afterward, at the nineteenth hole. If you want to form a real bond, however, we believe it might be time to think beyond the golf course. So if you're looking for a different opportunity to share quality time with your team members, there's a broad palette to choose from. IDEOer Hilary Hoeber believes that preparing gourmet meals together (and then eating them, of course) is a great way to strengthen team bonds. My friend John Berger at G2market says he likes to take colleagues scuba diving with him because it builds both mutual trust and shared experience.

But Neil Grimmer and Chris Waugh at IDEO may have carved out the ultimate niche in the world of radical collaboration: They do triathlons with their favorite clients. Neil started the trend when he discovered that one of our clients at Mercedes was also a serious runner. The next thing we knew, Mercedes was sponsoring them both (complete with team T-shirts and other gear) in a triathlon in Germany, and Neil was doing early-morning training runs with Mercedes manager Manfred Dorn before every team meeting. Mercedes even scheduled the final presentation for the project to take place two days after the triathlon, so both companies could celebrate completing the event.

They may have carved out the ultimate niche in the world of radical collaboration: They do triathlons with their favorite clients.

Now the idea has a life of its own. Neil and Chris compete every year in the Life Time Fitness Triathlon in Minneapolis—sponsored by our client Cargill—and work on Cargill projects in between. More than a hundred Cargill employees turned out for last year's event (not all of them actually competing), and the joint effort has built a lot of goodwill between our firms.

So even if—like me—you're not quite prepared to go scuba diving or run a triathlon with the team, look for new creative options for building camaraderie that go beyond the ordinary.

Sweat Equity

Thomas Edison once famously claimed, "Genius is one percent inspiration and ninety-nine percent perspiration." But often the most important step in a radical collaboration is simply finding a way to work together in the first place.

Far too many companies pass up great opportunities to partner because at first blush they may not seem to provide ample incentives. Just as you need to be creative about the work, you need to be open-minded about how to make collaborations happen.

When the Boston Beer Company approached us a few years ago to rethink the classic tap handle, we figured our normal fees might be an issue for a small company updating an object that they actually give away. But we were so motivated to collaborate with the brewer of Sam Adams Boston Lager that we worked out an unusual arrangement. They paid us partly in—you guessed it—beer. Every time the good-natured Sam Adams team showed up at our local Boston-area office, they rolled in a handcart with half a dozen cases of beer. Every IDEO Boston party

The ultimate business collaboration: finishing a triathlon with your team members.

for the next two years featured Sam Adams, and many a bottle of their hearty lager followed team members home as well.

It may not have been our most profitable job, but it sure was fun. We certainly had no problem staffing the project among IDEO's young design team. And I suspect we were even more diligent than usual in our field research. We explored the "theater of beer," that magic moment in which the bartender serves your favorite brand and the tap is center stage. We noticed that nearly every brewery had chosen to emphasize tradition with a painted or varnished wooden tap handle reminiscent of nineteenth-century British pub gear. We felt that made sense for modern high-volume beers that needed to emphasize their connection back to the rich tradition of brewing. But we argued that an authentic craft beer like Sam Adams could be poured from a more contemporary tap.

By exploring new materials we opened up all kinds of possibilities. Using injection-molded plastics in the handle raised the possibility of introducing electronics and translucency. We brainstormed ideas like having the handle glow from an internal light source, tallying a running count of drafts poured, or generating a mini light show during the pour. In a lighter moment we even contemplated an electronic "lottery" that periodically delivers the jackpot of a free beer, like the handle on a Las Vegas slot machine. In the end, we kept it simple—and economical: a cool blue translucent handle and a stainless-steel shaft that stands out from every other tap handle in town. We gave Sam Adams Boston Lager a renewed role in the theater of beer and laid out a potential path for the future evolution of the tap handle icon. And the collaboration was considered so mutually rewarding that when the Boston Beer Company decided to highlight their multivaried seasonal beer with a changeable tap handle, we worked together again on that project and our fridge was well stocked once more with their smooth, full-bodied brews.

Pass That Baton

To my mind, there's nothing like the critical baton pass in track and field's thrilling 4 × 100-meter relay to bring home the importance of working in sync. As anyone who's run one can tell you, in a relay race the baton pass is more important than the time of any one individual leg, because if you blow the handoff, it's game over. No relay team in the history of the Olympics has ever dropped the baton and won the race. A failure of teamwork overshadows the achievement of even the most brilliant individual performance.

To those who are not track aficionados, the relay may seem to be simply a test of who can put together the four fastest men or women on the planet. But the reality is much more complex. The best teams carefully blend talent and resources. The leadoff runner must have an explosive start. The middle runners need to be world-class at both *receiving* and *passing* the baton. In the 4 × 100 relay, each runner completes one of the four 100-meter legs, though in practice they cover slightly different distances. They must pass the baton within a 20-meter exchange zone or be disqualified.

The laws of physics dictate that the baton passes be smooth and take place with minimal deceleration. If the second-leg runner can accelerate before receiving the baton, precious little momentum is lost. America's entrants for the men's 4 × 100-meter relay in Barcelona were

A graceful handoff combines speed, flexibility, and close collaboration.

Mike Marsh, Leroy Burrell, Dennis Mitchell, and the legendary Carl Lewis. Each of them runs the 100-meter dash in about 10 seconds, so you might guess that their combined time would be about 40 seconds, right? Sounds logical. Yet these four remarkable men, each running 100 meters and passing the baton three times, put together a combined performance of 37.4 seconds for a world record—averaging more than 26 miles an hour! But how is that possible? It's possible because at the moment starter Marsh executed a smooth-as-silk handoff to Burrell, his teammate was *already nearing top speed*. When the legendary Carl Lewis took the final handoff, he blew by the competition—hitting 28 miles an hour at the finish and helping his team take home Olympic gold.

Relays are won or lost in the handoffs. If the receiving runner starts too late, momentum is lost. Start too early and they might run out of the exchange zone before receiving the baton and be disqualified. We've all seen botched baton passes—on the track and in business. They're invariably due to a lack of coordination and communication. And how often is the handoff synchronized perfectly so that the recipient is already nearing top speed when they get the baton?

One baton pass can be the difference between success and defeat. At the 2004 Olympics in Greece, the U.S. men's and women's 4 × 100 teams each essentially gave a clinic on how poor baton work can slow or scuttle a strong team. Tired from her long jumps and new to the relay team, Marion Jones failed to make her handoff in the exchange zone, disqualifying her heavily favored team. Meanwhile, the U.S. men's team lost to Great Britain due mostly to one bad final exchange.

There may be no better example of the importance of smoothly passing the baton than the 1936 Berlin Olympics. Hitler was counting on the German women to prove their dominance in the 4 × 100-meter relay. Coming up on the anchor leg, the Germans led by a full seven meters, an almost insurmountable margin. But the German anchor dropped the baton and Elizabeth Robinson of Riverdale, California, safely passed her baton to the equally fleet Helen Stephens of Missouri, who won the gold medal in record time.

Passing the baton in a modern organization can be even trickier than in a relay, but the metaphor still applies. Success depends on pick

ing the right team and casting them in the proper roles. All participants strive to achieve their personal best while thinking of the team's
performance throughout. If you work on those exchanges to the point
where they become smooth and fast, you'll be amazed at how much
you can achieve together.

Long Distance

In today's global economy, your baton passes are just as likely to need
to cross over oceans as well as departments. Collaboration gets a whole
lot more challenging when your team is spread around the country—not
to mention the world. Increasingly, however, that's the nature of modern
business. It's a large part of why IDEO has split its business into practices, deemphasizing geography and recognizing our global capabilities.

Let me give you one small example. As I write, I am working in
California on a project for Bertelsmann, based in Gütersloh, Germany.
The account manager is a Brazilian working out of IDEO Munich. The
project leader, a Canadian, calls our London office home. And the proj-
ect is being done in Paris, organized in part by a trilingual woman from
Beijing.

So how do you pull off an international project? Start with some
genuine face time (videoconferencing doesn't count). Going out for coffee or lunch is how you make the personal connections that build the
kind of relationship to enable you to phone someone an ocean away
and ask for a favor ASAP. Humans are still hardwired to believe that
breaking bread with one another matters.

Once you've made that initial human connection, you can better
maintain the momentum by establishing multiple lines of communication. E-mail is not enough. At IDEO, we build e-rooms, virtual spaces
dedicated to projects carved out on the company's digital network. Team
members make and manipulate a project-specific Wiki (an extremely
malleable form of Web page). We also often do Web-enabled meetings,
where we are all looking at the same presentation or documents. We're
not in love with any one technology, but we are willing to adopt whatever tools increase the human bandwidth of team interaction.

When you can't manage face time, it's critical to make a space for what some might consider the soft stuff—bio sketches of team members, insights about their hobbies. When you're working in close proximity, you're likely to glean this through osmosis, such as when you ask, "What did you do over the weekend?" Leave room to gauge your team's mood and state of mind. It's hard to pick up on a team member's simmering frustration if e-mail is your only form of communication. Street-savvy Collaborators know never to confuse an e-mail with genuine human communication. I'm constantly dumbfounded at how many people send out e-mails when what they really need to do is to make a far more efficient two-minute phone call. That's maybe the most important suggestion. It's even more critical in coordinating global innovation efforts. Make time for lots of little and long conversations. Don't say something in an e-mail that you wouldn't say in person. Avoid ambiguous e-mails that might possibly be misinterpreted by someone who's had a bad day. And if possible, never initiate contact through an e-mail.

Whole Teams

Just down the street from IDEO's Palo Alto campus is a Whole Foods supermarket, so I've experienced the upscale-food-store phenomenon firsthand—and frequently—for the past several years. It's the closest place to grab a quick lunch near the office, and if you get a craving for an espresso chocolate-chip cookie in midafternoon, Whole Foods is always more than happy to help you out. When IDEO prototyped a new grocery shopping cart for Ted Koppel's *Nightline* show a few years back, the store was an easy first choice for where to "market-test" our concepts.

Whole Foods' rise is more than simply the tale of a well-designed store or the growing popularity of wellness and healthier foods. By challenging conventional wisdom about labor and management, the chain appears to be doing more than selling a different class of food. It's literally offering a new way to manage a labor-intensive business.

Whole Foods' financial results are remarkable. In an industry plagued by razor thin margins, the company earned nearly $200 million in profit over the last two years. Meanwhile, its closest competitor, with seven times the number of stores, couldn't match Whole Foods' profits, and others were losing money in a tough market. One of the secrets to Whole Foods' recipe for success is the collaborative model that permeates their operation. While the big chains have plenty of managers and clerks, Whole Foods generates more creative, engaged, project-oriented teams.

Each store has eight in-house teams, and each team does its own hiring. New employees are hired by the bakery team, the seafood team, or whatever group they're joining. A month after you're hired, two-thirds of your fellow employees must vote to keep you on. In other words, you've got to pull your weight. And like the best project teams everywhere, the Whole Foods functional groups have a lot of say in everything from what they stock in their area to how they display the food. Increased sales and profits for a team translate into extra compensation for members.

Adding to the overall team spirit, there's a sense of equality and transparency in store management that's rare for any organization. For example, no executive can make more than fourteen times a store's lowest wage-earner, and every employee can consult a book that lists everyone's pay. More important, the company has reams of data on sales, volume, and finances that teams regularly consult as they gauge their own success. Everyone is eligible for stock options. There's even a "Declaration of Interdependence," which along with laying out the company's approach to food and its customers clearly articulates how it feels about its employees. Whole Foods claims it has no room for "us-versus-them" thinking. Active, broad participation is its continuous mantra. Some of the ways the company says it achieves those things are worth repeating:

○ Self-directed teams that meet regularly to discuss issues, solve problems, and appreciate each other's contributions.

○ Increased communication through Team Member Forums and Advisory Groups, and open-book, open-door, and open-people practices.

○ Commitment to make our jobs more fun by combining work and play, and through friendly competition to improve our stores.

○ Continuous-learning opportunities about company values, food, nutrition, and job skills.

In other words, teams are the backbone of Whole Foods. And the lifeblood of its operations is collaboration. Chances are if you talk to other Whole Foods regulars, you'll hear similar stories. Whole Foods workers go out of their way to help customers find products and answer questions about everything from a great fish sauce to how to prepare a certain dish. They're generally in good spirits and eager to help. This shouldn't be any surprise. They're part of a close-knit group. As far as they're concerned, they're on a winning team.

To me the lesson for Collaborators is simple: Transform the work of your organization into projects headed by teams. Give them a powerful role in their work. My experience tells me you're bound to reap positive results. Who knows? You just might even pump profit into a traditional business that the pundits gave up on long ago.

Teamwork: The Soccer Model

As a father, I've watched youth soccer, still relatively new to America, rapidly gain ground on football. During my childhood in Ohio, we played football in the backyard from late summer till the first winter cold snap. I hardly ever saw a soccer ball until I went away to college. Now it's almost exactly the opposite. I'm not even sure my ten-year-old can throw a decent spiral, but he's got a pretty wicked soccer kick and three seasons under his belt. For his generation, soccer has become the most popular fall sport for boys and girls (some 21 million American children and adults now play it). What, you may be wondering, does this have to do with teamwork in the business world?

Soccer is a team endeavor in the truest sense and a powerful international metaphor for collaborative enterprises of all types. The best soccer coaches do virtually all of their coaching in practice. Steve

Negoesco, coach of the nearby University of San Francisco Dons, recently retired with a record of 544 victories, the most ever for an intercollegiate soccer coach—and four national titles. Negoesco's secret? He chose a captain before the kickoff and then sat up in the stands where he could get a good look at the game. He didn't pull a player who wasn't performing in the first twenty minutes or so. Building confidence, he felt, was more important than winning games. If somebody was clearly off form, he might make a change at halftime. If someone was injured, the team captain—not Negoesco—waved in a replacement.

What did his amazing organizational patience and discipline achieve? His players weren't constantly second-guessed from the sidelines. They felt a great responsibility to the team. They earned their starting roles and meshed together as a team with remarkable creativity, skill, and coordination. Negoesco coached his teams so well in practice that by game time each player was himself a coach, ready to tackle any number of challenges on the field. They could think for themselves while playing as a team. Couldn't you use more people like that—strong, responsible, and creative individuals ready to exploit diverse opportunities?

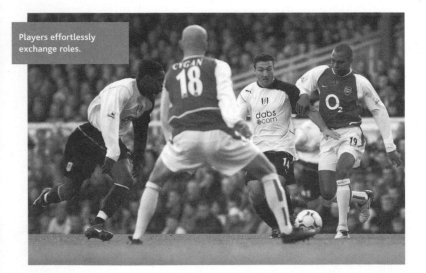

Players effortlessly exchange roles.

Soccer teams are a lot like good company teams. Each must master basic skills and constantly handle and pass the "ball" down the field. The best teams cherish diversity and have players who are short, fast, tall, or strong. Individuals who enrich the team. Soccer consists of a series of interlinking passes, like project-team handoffs. Watch a professional game and you'll notice that good players are constantly running to form and re-form triangles all over the field, giving the ball handler angles and passing options, just as team members engage in a project in a variety of ways.

Rather than focusing on the specialized skills of a specific position, as in football, soccer requires overlapping, the teamwide blending of skills and responsibilities. We use the same kind of thinking at IDEO. We've increasingly come to recognize that just because someone's degree is in engineering doesn't mean they can't contribute in a creative brainstorm about a new service. Conversely, we ask our designers for their insights on challenges that might seem to lean toward engineering solutions. Good teams ask people to stretch and cover for one another, to fill the gaps.

When Dutch teams began playing the remarkably innovative—and eventually much imitated—brand of soccer known as "Total Football," it required rapidly changing roles and instantaneous communication among players. A defender who won the ball from an opposing forward might sprint down the wing, morphing into a forward. In response, a midfielder would swing back to cover the defender's position to guard against a counterattack. More efficient teamwork spawned greater cohesiveness—and more goals. It made for more exciting soccer—and a winning team.

That same kind of real-time flexibility and broad-based communication on your team can help reinvent how you approach projects and rack up "goals." Here are some of the keys we've highlighted that can help build better teams:

○ **Coach More, Direct Less**
Good executives and managers inspire their staffs to develop their confidence and skills so they can seize critical "big game" opportunities.

○ **Celebrate Passing**
Break teams into smaller groups of three to six to increase the number of triangles where team members can pass ideas and responsibilities.

○ **Everybody Touches the Ball**
Find one or more key responsibilities for every player. Don't relegate team members to the corporate equivalent of football linemen who rarely touch the ball.

○ **Teach Overlapping Skills**
Create opportunities for team members to assume nontraditional roles and push forward initiatives. Invite techies to brainstorm big concepts and sketch out ideas and encourage those with a less technical bent to access technology issues. Find out team members' unique passions and interests, and put them to work.

○ **Less Dribbling, More Goals**
Encourage the sharing of ideas and initiatives. Solo dribbling can give a project the critical first push, but then you need teamwork to bring a project home.

Sharing the Journey

The ups and downs of a collaboration are part of what makes it all work. As business guru Gary Hamel says, the process of collaboration can in some cases be more important than the finished product itself. To illustrate his point, he uses a metaphor from the movies. If you walk in during the last five minutes of a Hugh Grant (or Cary Grant) movie, the final scene is likely to be the same—the handsome star is in the arms of some leading actress, buoyed by a swell of romantic music and the supporting cast. Out of context, you may be unmoved or unimpressed by that tender scene. But if you sat through the whole story, watching the roller-coaster romance go through breakups and misunderstandings while building up to that climactic final moment, the same ending might be poignant and touching. It might even bring a tear to your eye. Your response is not about the result (he's always going to get the girl), it's about the process, the journey, the shared experience.

The same is true of business collaborations. Going through the process together builds understanding, commitment, energy, and momentum. That said, collaboration is never as smooth as the fiction portrayed in the movies. Divorce rates are a sad reminder of that truth, as are the number of business partnerships that hit the shoals. So what do you do when your project bumps up against these natural speed bumps? Consider an approach that may sound a little counterintuitive: Embrace your critics.

Co-opting Your Opponents

Take, for example, this story told by Maura Shea of our London office. Maura happened to be working on a project for a well-known U.S. hospital, looking for innovative ways to accommodate patients and their families in a new space. It was a challenge. Many of the top officials and doctors were understandably skeptical of a new "consultant" hanging around (we shy away from that title). As Maura put it, they'd just seen so many failed initiatives that they didn't believe change was possible. They were suffering from "consultant fatigue." Part of the problem was that many of these people were essentially pulling double shifts, doing their regular jobs while working on the project. Many were embittered, to put it mildly.

The doctors, especially, were intensely skeptical. One day Maura was walking down the hall when a doctor angrily shouted at her in front of several people, "How expensive are you? What are your rates?" Maura held her tongue. The doctor went on, "We've been waiting years for something to change around here! How are you guys going to do anything different?"

Maura still didn't lose her cool. "I hope you will join the project," she told the doctor. "Without your participation, we won't be as strong a team."

Making that initial peace offering took a lot of strength and self-assurance. But Maura meant what she said. The doctor had strong opinions and doubtless strong ideas about how things might be changed. So Maura began trailing Doctor X around the hospital, watching how he worked and engaged the staff and patients. He was in many ways a great subject: smart, opinionated, eager to voice what did and did not work for him on the project.

It's a wonderful lesson in co-opting your opposition. Instead of being offended by their arguments, why not listen and respond to their concerns? They often have valid points. The payoff can be extraordinary. There's nothing like the conviction of a convert to boost team momentum.

On another collaborative project with an architectural firm, a senior partner at the firm told our team up front that we were wasting his time. In turn, we requested that he give us a chance and join our team. Not only did he embrace the process, but midway into the work, he asked if he could write a case study about our process for an architectural magazine. And just like that, a onetime critic morphed into a passionate advocate.

Whenever you collaborate, you're likely to meet opposition. The best advice I can give to anyone playing the role of Collaborator is to be patient. The next time your efforts are met by a sincere critic, squint your eyes and imagine just what a force this person could be if he or she were on your side. By listening to their concerns and complaints, you'll already have taken the first and most critical step in ultimately winning them over.

CHAPTER 6
The Director

I dream for a living.   —FILM DIRECTOR STEVEN SPIELBERG

Since the publication of *The Art of Innovation*, I've been delighted to note a popular trend in the vocabulary of business organizations. Hundreds of corporations around the world have recognized the seminal role of innovation, creating positions for vice presidents of innovation, chief innovation officers, and, yes, *Directors* of innovation. The positions and titles are a tacit recognition that just like the chief financial officer or chief operations officer, the job of orchestrating a company's innovation efforts is critical to the long-term health of the enterprise.

You know the Director. She's the person mapping out the production, crafting the scenes, bringing out the best among actors and actresses, honing the project or company theme, building the chemistry, getting it done. Our firm has had the good fortune to work with Ivy Ross, a talented Director who oversaw a tremendous innovation effort at Mattel dubbed "Platypus." She commandeered a large project space on the corporate campus, pulled top Mattel designers and project leaders away from their day jobs for twelve weeks, and plunged them into vigorous prototyping, rich observations, and all sorts of unusual activities. They participated in some great group improv sessions. They shot our favorite foam Finger Blaster rockets at one another regularly. Amid all the fun, they were gathering insights that pushed the project forward.

By week eleven, some Mattel executives were losing patience. But they put away their doubts when that very first Platypus team created a new girls' toy platform for Mattel called Ello (as in "say hello to Ello"), which racked up more than $100 million in sales during its launch year.

An innovation-oriented Director helped the Mattel Platypus team create a whole new product platform in just three months.

Choose Your Flavor

There's more than one Directorial style at IDEO. We take an independent filmmaker's approach—so there's room for all sorts of Directors, from the calm confidence of Sofia Coppola to the frenetic energy of the Coen brothers. IDEO's Bill Moggridge, for example, succeeds with his own personal style. A few years ago, he created our Know How speakers series from scratch with a light touch—passing stewardship for organizing our Thursday-afternoon sessions to the many individual IDEO "curators" who liked Bill so much that they gladly helped recruit speakers such as Malcolm Gladwell, Jeff Hawkins, and Stephen Denning. Moggridge single-handedly disproved Machiavelli's dictum that "it is better to be feared than be loved." He used his personal warmth to get people to do things for him.

I've learned a lot about the Director role myself from watching my brother David in action for the past twenty-seven years at IDEO. Contagious enthusiasm might best describe David's style. He is really good at getting people to take intelligent chances and giving them the opportunity to recover from their failures. He's inspired countless IDEOers,

Stanford's new design institute applies "design thinking" to the world of business.

and now he is doing the same at Stanford, where he's creating a new design institute (which some people are already calling the "d.school" to contrast it with Stanford's prestigious Graduate School of Business— the "B-school").

Unlike the designers portrayed in the movies, outsized egos are relatively hard to find at IDEO, partly because of the tone set by my brother and his chosen successor, CEO Tim Brown. Tim also projects an understated confidence that leaves ample room for the achievements of our other leaders. He has renewed our processes of innovation by centering our efforts around market-focused practices.

One standout Director among the clients I've met in the past few years is Claudia Kotchka, Vice President of Design, Innovation, and Strategy at Procter & Gamble, who was recently described in *Fortune* magazine as "the most powerful design executive in the country." Reporting directly to CEO A. G. Lafley, Claudia is part of P&G's new secret formula for success. She's adept at playing the Anthropologist and Collaborator roles, but her true calling is as a Director. Like many people in that role, Claudia relies on far more than just her formal authority. She uses sheer force of personality to coax, cajole, and corral people into seeing things her way. Among many other initiatives, she helped set up a corporate innovation fund and then asked managers in P&G's global business units to suggest "problems worth solving"—the kind of things that kept them up at night. She rejected 90 percent of the proposals, mostly because she considered them "not hard enough," but ended up with a great list of projects to move forward on collectively. We think P&G has created an innovation-driven strategy that's bound to produce results.

FIVE TRAITS OF SUCCESSFUL DIRECTORS

1 **They give center stage to others.**
Directors are content to let others take the spotlight, confident in the knowledge that their behind-the-scenes work will make the whole production come together. (You never saw Frank Capra on-screen—except at the Oscars.)

2 **They love finding new projects.**
Directors gladly step up and lead when the need arises, and they view team chemistry as an intrinsic part of project success. They put together the very best team they can find and afford, and are sometimes willing to delay or restructure the project to accommodate the right players.

3 **They rise to tough challenges.**
The history of moviemaking—like the history of enterprise—is filled with long hours, ebbing budgets, looming deadlines, and inevitable setbacks. Directors expect hardships along the way and are up to the task.

4 **They shoot for the moon.**
Directors pursue bold strokes and lay out goals that seem difficult or even impossible to achieve. Then they work to make their dreams a reality.

5 **They wield a large toolbox.**
Solving problems in real time, Directors are ready to improvise with whatever techniques, strategies, and resources are at their disposal. Directors embrace the unexpected.

In Hollywood, Steven Spielberg is the epitome of someone who directs his cast and crew to capture the audience's imagination, while in business that role falls to people like Steve Jobs, who has proved himself brilliant at motivating his teams to create something "insanely great." Both Directors have the ability to bring out the best in their team, often with a contagious enthusiasm that spurs individuals and project groups to extraordinary achievements.

The role of Director is more complex and nuanced than any other in the world of innovation. That said, it's important to start with a few basic truths. The Director's first and most prominent task is to always keep the production moving forward in the general direction of the goal. Directors need to have a fierce grasp of an elemental truth of their business— whether they are making movies or delivering services or creating customer experiences. You are not just in charge of today's operations. You are responsible for making sure there is a tomorrow. You must constantly juggle the balls necessary to make new projects happen, to replace a just-finished innovation with a fresh exploration of another opportunity.

> There's an old adage in Hollywood that "directing is 90 percent casting." Great Directors build a team of people who need little direction and can lead by example themselves.

Of course, the most critical part of directing is getting started: getting the work, spawning a creative culture, and hatching the ideas. Directors are unlike all the other personas because their main purpose is to inspire and direct other people, developing chemistry in teams, targeting strategic opportunities, and generating innovation momentum.

There's an old adage in Hollywood that "directing is 90 percent casting." Great Directors build a team of people who need little direction and can lead by example themselves. Directors can make something out of nothing. They can field a project team and motivate its members, even without formal authority.

Jump-Starting Innovation

At IDEO, David Haygood sometimes takes on the Director role. Once you get him on board, he'll create his own energy. Haygood believes the catalyst for innovation projects is face time with the key players. He and his team have a knack for getting us in front of top executives of major corporations. Of course, the best collaborations and partnerships happen when the key decision-makers are on the same page, but as we all know, it's no simple matter to set up such meetings. Traditional approaches—like calling out of the blue or dashing off an e-mail—often don't succeed. But Directors make things happen.

Not long ago, David had a brief introductory meeting with a very senior executive at a car manufacturer. It went well, and the company hired IDEO to do a pilot project on some new interior concepts. From an offhand comment after the meeting, Haygood learned that the exec was headed to Las Vegas the following week for an automotive conference. That kind of small detail might have gone in one ear and out the other. But Haygood sensed an opportunity.

He felt that what the client relationship really needed was for our senior leadership to talk in person with the auto executive. But such meetings can take months to schedule. So the following week, Haygood went way out on a limb and convinced our CEO Tim Brown to catch a flight with him to Vegas. Haygood had no appointment and no fixed plan. He risked embarrassing not only himself but also the head of the firm. (Early in my career, I watched my mentor at General Electric get fired for a similar offense.) In fact, since the whole conference was closed to those outside the automotive industry, David and Tim might never have gotten in the door.

Through a combination of serendipity and perseverance, Haygood and Tim Brown did manage to bump into exactly the person they were looking for, just as he emerged from a conference call. The executive seemed genuinely pleased to see Haygood and suggested that they all have a drink at the company's hospitality suite. Since he hadn't been to the hospitality suite himself yet, ironically it was David Haygood who guided him there. The impromptu meeting went off without a hitch and helped build the relationship between our companies. No, it wasn't business as usual, but that isn't how many of the best projects take hold. The fortuitous meeting jump-started the collaboration. And the auto exec was happy to get a chance to meet Tim and hear his perspectives on the work firsthand.

I'm not suggesting you make a habit of flying to distant cities to randomly bump into key partners. But there's a lot to be said for David Haygood's gumption. Everyone knows that simply mailing a formal résumé almost always fails. So does responding to a traditional Request for Proposal. When seeking new business or nurturing a new relationship, you've got to be more creative about how you approach getting quality time with the key players. Think in terms of opportunities. I doubt David Haygood loses much sleep worrying about failing. That may be the first step to making a great first impression—and the deeper connections that can help launch a project.

Getting Started with a Brainstorm

When I'm speaking to business audiences, whether it's in Europe, Asia, or America, one of the most frequently asked questions I get about innovation is "Where should we start?" The innovation challenge can seem so complex and ambiguous that sometimes organizations have trouble kicking off the journey. If you find yourself in the Director role trying to get the ball rolling, I believe one of the easiest ways to get a quick return on innovation is to set off a chain reaction of brainstorming throughout your organization.

Why brainstorming? It's high-energy. It's fun. And it can boost morale and generate results faster than any other technique I know of. Be sure to keep the initial barriers to entry low, and start with problems you think you can make a dent in right away. Build a culture of brainstorming, and you've got a great start toward nurturing a culture of innovation.

So here's a simple program to try with your organization. Launch a series of brainstorms on different topics, anything from how to reduce your customers' waiting time to how to make more effective use of an empty office space. The precise topics are not critical at first, since your initial purpose is simply to increase the rate of ideation. Later, you can tackle the harder brainstorming topics.

You could sponsor lunchtime brainstorms once a week, or every payday (when people are in an especially upbeat mood), for the next six months. Recruit a master of ceremonies for each session—someone brimming with self-confidence and energy. Next, round up five to ten interested participants (mixing in at least a couple of new people for each session). Look for staffers sporting an eclectic mix of backgrounds, preferably skewed initially toward more outgoing people with nimble minds. Offer pizza or sandwiches, with indulgent "reward" food like M&M's or chocolate-chip cookies. (I'm all for healthy food, but sometimes a fresh plate of cookies really does add fuel to the fire of a good brainstorm.) Also, stock the room with plenty of colorful marker pens, Post-its, and other materials for capturing the ideas as they bubble up.

If you head the team or the organization, make it abundantly clear to everyone that the new program has your enthusiastic support. I would encourage you to attend the first couple of minutes of some sessions, when the brainstorming topic is introduced. Then—and I think this is pivotal early on when brainstorming is at a fragile, embryonic state—get out of the way. Tempting as it is to stay and share your knowledge of the business, I believe in many, many cases the presence of a senior executive or CEO will be counterproductive.

Whatever your role in your organization, check out the sidebar on brainstorming for a quick review of some of the basic brainstorming principles I cover in more detail in *The Art of Innovation*. And remember, practice makes perfect. There may be some rough spots at first. Occasionally, you'll encounter the dreaded "dead air" moments, where the whole room goes quiet, sucking energy out of the group ideation session. You'll likely learn that some traditional managers are not really naturals at running brainstorms, while others have a latent talent for it. Gradually, you'll find out who can help draw out the best ideas (and who has trouble deferring judgment). A core of two or three good, strong masters of ceremonies will keep the good ideas flowing—pay-day after payday.

Here's the unexpected return on this new brainstorming program: The benefits go beyond the immediate ideas you generate. Regular brainstorming is as critical to an organization as regular exercise is to your health. It creates a responsive, innovative culture. Stanford professor Bob Sutton has tracked several distinct benefits of regular brainstorming for organizations. Here are my three favorites:

○ **Brainstorms Support Your Organizational Memory**
You can't always afford to have the most experienced members of the organization on your project team, but you can invite them to a one-hour brainstorm and tap into their organizational knowledge and expertise. Through brainstorming, the group explores possible solutions from the past, present, and future, which constantly engages the memory and intelligence of your company. Fluency in brainstorming skills can translate into heightened responsiveness, an increased ability to adapt to and anticipate new challenges.

Brainstorming helps fuel your engine of innovation.

○ **Brainstorms Reinforce an Attitude of Wisdom**

As mentioned in Chapter 3, an attitude of wisdom is a healthy balance between confidence in what you know and a willingness to listen to ideas that challenge your worldview. Participating in frequent brainstorms forces you to match wits with other creative minds and realize that others have ideas that improve upon your own. The process can be humbling sometimes, but it makes you wiser.

○ **Brainstorms Create Status Auctions**

Not only are creative brainstorms exciting and fun, their free-spirited atmosphere gives all kinds of people a chance to shine. In a multidisciplinary brainstorm, you get a chance to see other people in action—and be seen yourself. Creative people who excel in the intense environment of a brainstorm can gain attention and status that might otherwise pass them by. Brainstorms reinforce their personal brand.

SEVEN SECRETS TO BRAINSTORMING

Here are a few highlights on how to tune up your brainstorms:

1 Sharpen Your Focus

Begin with a clear statement of the problem, a question that's open-ended but not too broad. Focusing on a specific latent customer need or one step of the customer journey can often spark a good ideation session. For example, "How could we gain deeper insights into the experience of our first-time customers?" would be a useful brainstorming topic for many organizations.

2 Mind the Playground Rules

We've stenciled our brainstorming rules high on the walls of many of our conference rooms: Go for Quantity, Encourage Wild Ideas, Be Visual, Defer Judgment, One Conversation at a Time. Even in a rule-averse culture, we've found these basic principles to be both instructive and empowering.

3   Number Your Ideas
Numbering your ideas motivates participants, sets a pace, and adds a little structure. A hundred ideas per hour is usually a sign of a good, fluid brainstorm, and even if the group is nearly out of steam when you hit number ninety-four, it's human nature to want to push on for at least half a dozen more.

4   Jump and Build
Even the best brainstorms hit plateaus. You have a flurry of ideas and then they start to get repetitive or peter out. That's when the facilitator may need to suggest switching gears: "How might we apply these ideas to . . . ?" Push forward with a small variation or cycle back to a promising earlier idea to maintain momentum and build energy.

5   Remember to Use the Space
Leverage the physical environment to make your brainstorm more effective. Let your brainstorm literally take shape and fill the room— write and draw your concepts with markers on giant Post-its stuck to every vertical surface. Capture your ideas in visual, low-tech mediums that everyone can share. Spatial memory is a powerful force you can use to guide the participants back on track.

6   Stretch First
Ask attendees to do a little homework on the subject the night before. Play a zippy word game to clear the mind and set aside everyday distractions. Borrowing from the world of improv, we often start with some form of warm-up, like free association, where I toss out a word or idea and another person quickly builds on it and tosses it to someone else. Athletes stretch. So do brainstormers.

7   Get Physical
At IDEO, we keep foam core, tubing, duct tape, hot-melt glue guns, and other prototyping basics on hand to sketch, diagram, and make models. Some of our best brainstorms have quickly leaped to roughing out an idea with a crude prototype.

Start with Some "Zip"

Brainstorming is just one of many tools in the Director's toolbox. Another opportunity sometimes overlooked or underestimated is the power of a name. Lending your team, project, or proposed product an active, colorful name can give it a boost. The best project names preserve the confidentiality of the work while giving the team something iconic to latch on to. At IDEO, for example, the Palm V project was code-named "Razor" to remind everyone that we were aiming for something slim and elegant. "Greenhouse" was a project to help a multinational company build renewal into its brand, and "Hamachi" helped a Japanese company launch into a fresh new market. And here's a hint: Most good project names lend themselves pretty readily to a distinctive team T-shirt that builds solidarity.

The naming process—whether you're naming a project, a product, or even a book—rarely happens in an instant. Often you have to come up with dozens of rejects before you settle on a winner. Sometimes it helps to talk over or write down the essence of the experience you're hoping to create, then see whether the name supports—or gets in the way of—the core idea.

History tells us that great names sell. No one much remembers the "Hookless Fastener," but when Goodrich transformed it into the Zipper in the 1920s, its success was practically ensured. As the company myth goes, Goodrich's president reportedly said, "What we need is an action word . . . something that will dramatize the way the thing zips. Why not call it the Zipper!"

Sometimes even an unlikely name can add to success—if it brings its own energy. Not far from where I grew up in Ohio is a family-run business named the J. M. Smucker Company, most famous for peanut butter and jelly. The folksy company advertised for years with the self-mocking slogan "With a name like Smucker's it *has* to be good." But in fact, the name has a pleasing quirkiness. It's certainly distinctive, and it even borders on onomatopoeia—suggesting that smacking noise your tongue makes after peanut butter has momentarily stuck to the roof of your mouth. For Americans (the biggest peanut butter fans in the world), the name—and the sound—is reminiscent of some of the nicer memories of childhood.

And as for unlikely names, there's a bike race in California that shows you can't always trust your intuition about what will work in the marketplace and what won't. Twenty years ago, this grueling annual ride (129 miles and four mountain passes) had a limited appeal to only the elite few willing to brave the strenuous course. But then the organizers renamed it the "Death Ride," and it enjoyed a surge of popularity. Apparently, the grim name appeals to the tough-as-nails psyche of the long-distance cyclist. Now each year something like 6,000 applicants vie for "only" 2,600 spots available at the event. It's doubtful that you have something you want to call the "Death Ride," but if a quirky, distinctive name could get people waiting in line for a chance to participate in what you have to offer, it's worth investing some time and money in the search.

Over the years, I've run into some people who make their living coming up with names, and I find their work to be both subtle and fascinating. Marc Hershon, a screenwriter and principal in the Sausalito branding firm Simmer, was a key member of the Lexicon Branding team that named the BlackBerry, the Swiffer, the PowerBook, the Pentium, OnStar, and the Subaru Outback. "Names are equal part science, art, and inspiration," says Marc, who has helped me name a few things. "It's a weird cross between improv and poetry—albeit very short poetry."

> Consumers can tell you that a car named Viper is faster than a Lumina before any other part of the car is designed.

And it turns out that not only do names carry great meaning, but even individual letters within the name have culturally specific connotations. For example, professional namers like Lexicon founder David Placek would tell you that consonants like *v* and *z* connote speed, while letters from the middle of the alphabet like *l*, *m*, and *n* are slower, more comfortable sounds. Consumers can tell you that a car named Viper is faster than a Lumina before any other part of the car is designed.

Names can make a huge difference in almost any new product or service. We believe that anything worth working on is worth naming, so besides the project names already mentioned, there is a growing list of IDEO places (e.g., the Lookout and the Grassy Knoll), programs (Know How, Innovation Seed Fund, etc.), and even roles. In the past twenty-seven years, IDEO has worked on some products with great names—the Nike Magneto high-performance shades, the ergonomic Steelcase Leap chair, a remote-controlled underwater camera called the Spyfish, the Polaroid iZone camera, and even a durable low-cost water pump called the Money Maker for use in developing countries.

Good Directors put fun and energy into naming your team, your project, your promising new service. Don't settle for a flat, business-as-usual description. Experiment and prototype with distinctive names. Have fun with them, and be on the lookout for names with surprising energy. Even today, a name like the "hookless fastener" wouldn't get you very far.

Try conjuring up a name with zip.

No Time Like the Present

IDEO has the luxury of having an abundance mentality, living in a world where demand for our services usually exceeds supply. During the recession of 2000, however, Tim Billing, a leader of our Palo Alto office, noticed that the team was running low on work. In most professional services firms, the senior partners are the ones most responsible for nurturing new-client relationships, but Tim told his team about the shortfall and asked them to keep their eyes open for a chance to help out. Pascal Sobol, a twenty-four-year-old engineer working for Tim, had never been involved in any of the firm's business development but he found a way to help anyhow. Pascal is a native of Germany, and on his lunch hour that day, he dropped in at Daimler-Chrysler's northern California research and technology center—located just a couple of miles from our Palo Alto office. Pascal struck up a conversation with the receptionist in German, chatting about how he'd grown up in Stuttgart. He understood Daimler had an overnight-courier package service to Germany. His dad used to work for Daimler. Would they mind adding a little package to the courier for his father?

The first impromptu visit led to another, and pretty soon Pascal got to know a few people at the research center. "Pascal came back one day," recalls Billing, "and said, 'Hey, I've set up a meeting with Daimler Benz.'" We made a presentation to the management team there and were hired by Daimler for two modest projects that led to more substantial work. All because a junior engineer barely out of college was confident and clever enough to put his German roots to use in striking up a connection. Pascal certainly didn't ask permission. The first time his boss heard about his lunchtime visit to Daimler was when he told him he'd arranged for a meeting with the prestigious multibillion-dollar corporation.

Many companies are fortunate to have at least a few Pascals. The larger question is whether the Director in your firm really gives them a chance to prove their mettle. Do you reward initiative? What might you do to encourage productive risk-taking?

A Marketplace for Talent

In project-based organizations like ours, one critical task for Directors is to allocate available resources to project teams. Matching talent and personalities to create the right team chemistry is tricky, but no process may be more vital to the health and vitality of your organization. For many firms, the staffing or resource meeting is where individual careers rise or fall—where the deft blending of individuals can make projects fly. Plenty of good ideas may be floating around your enterprise. Here's where you invest the human capital necessary to turn them into reality.

I'd like to say we mastered this task long ago, but we still consider our weekly resource meeting to be a work in progress. Here are a few tactical elements that we've found can help make such meetings more successful. First, pick an open area to hold the meeting, as opposed to a conference room—you're likely to have broader participation and strike a less corporate tone. An open-air setting adds transparency to the process, which reduces anxiety among the team and helps build support for the resourcing decisions that get made. We've also noticed that it's easier to keep the meetings from becoming stale by having two or even three leaders. There's the conductor, who keeps the agenda on track. There's the organizer or scribe, who records the staffing decisions. Finally, there's the referee, who promotes a positive spirit and sorts out disputes when they arise.

Rich, varied presentation mediums are also important. We use a combination of whiteboards, posters, bulletin boards, and projection screens to create a more interactive environment. Someday a fully digital solution will feel this interactive (like watching Tom Cruise orchestrate data on huge screens in *Minority Report*), but at the current state of the art, we find that mixing physical tools and computer-based systems yields a more visible and group-oriented process.

Smart companies avoid defining their staff too narrowly when matching people to tasks. Many people boast diverse talents and skills apart from their formal training. It's important to look for and leverage those complementary gifts where possible. Knowing that someone is a photographer by avocation, for example, might sway us to put them on a project where they could build an image database to help tell the story afterward. Or knowing that someone is an avid cyclist might make us more likely to put them on a wellness-related project team.

At every step, we try to personalize the process, keeping in mind that we are influencing lives and careers, not just fitting people into the next empty slot. Resourcing guru Lisa Spencer had the clever idea one day of putting employee photos (from our intranet site) onto magnets, and suddenly meeting leaders could literally grab and slide a person's image onto one of the projects sketched out on the whiteboard. Along with each person's photo, the magnet lists their name, discipline, location, and extension. The pictures add another visual dimension to the weekly marketplace for talent. Teams are all about chemistry. Pictures help you begin to see a good team in the making.

In a nimble and rapidly changing organization, every resourcing meeting is different. Some sail along, and some get bogged down on a single project. It helps to consider the meeting a prototype, to periodically ask what is working and what is not. You've got to advance individual careers while weighing the best interests of your company. Balancing the elusive goal of matching talent to projects is no easy task, but the process is very rewarding when you get it just right.

SET EXPECTATIONS

Before embarking on a continuous-innovation program, it's crucial to set expectations. How committed will your organization be to the process? Here are seven questions we've found that help companies prepare to launch new efforts. Answering them will help ready you for the hard work ahead.

○ How will your company define a successful innovation program?
○ How will your organization fund the innovation process?
○ What corporate resources (staff, space, technology, etc.) will be available to support your effort?
○ How often will the stakeholder groups meet to review your innovation propositions?
○ How many task teams will you sponsor yearly? How often will you put together these teams?
○ How much logistical support will be given to your innovation staff (time away from regular duties, prototyping tools, administrative support, etc.)?
○ What rewards or recognition can people expect for participating in this program?

Pay attention to how you answer these questions. If it's a struggle to answer them, your innovation program may lack the organizational and logistical support it needs. Good Directors give their innovation teams every edge they can. Lobby your firm's management to gain the financial and corporate backing that can help make your innovation programs a success.

Targeting Opportunities

Directors wisely marshal resources. Sometimes there's a tendency to believe that certain challenges are beyond the capabilities or resources of a team or company. That's part of the magic of innovation. Target the right part of a problem and you can leverage your investment. More is possible than you might imagine.

Consider this story of Brazil, a country that brings to mind images of glorious beaches, stupendous soccer players, and vast jungles. World-leading science—not to mention commercially viable science— isn't what one normally thinks about when it comes to the vibrant South American nation. Few would have guessed that Brazil would undertake a bold government and industry effort to become a world-leading center of genome research. But in the late 1990s, the government wisely opted to focus its early biotech research on a manageable but important challenge. Brazilian researchers chose the perfect initial target—a relatively simple bacterium that had the dual attributes of containing only 3,000 genes and being responsible for major damage to Brazil's vast orange crops. Brazil promptly sequenced the strain of *Xylella fastidiosa,* leading to yet more breakthroughs. Another strain of the bacterium was sequenced—a pest that threatens the California wine industry.

The rapid breakthroughs led to Brazil developing an innovative decoding strategy remarkably different from that practiced in most industrial nations—"Open Reading Frame Expressed-Sequence Tags." Soon Brazil cracked the 80,000 DNA sequence of the sugarcane— another huge national crop—and then, in early 2002, the researchers set out to master the almighty coffee bean.

As java lovers may have heard, in late 2004 Brazil cracked the genetic code for the coffee bean. "This is a jump of at least two decades in the race to unlock the coffee genome," Brazilian Agriculture Minister Roberto Rodrigues proudly announced at the time, heralding the creation of a "supercoffee" that will take the world by storm.

There's no doubt this was a big innovation for Brazil. The nation's $3.3 billion coffee industry employs 8.5 million Brazilians. Interestingly, Brazil has banned the planting and sale of genetically modified crops. But by identifying 35,000 coffee genes responsible for everything from aroma to caffeine, vitamins, and flavor, they will be "cross-pollinating" plants to breed superior varieties. They'll also have a jump on the rest of the world.

By starting modestly—in sequencing the DNA of orange and vine pests—Brazil has gained critical DNA momentum. It's a great argument for the power of a good Director, starting humbly but building up to major innovation opportunities.

What's the coffee-bean challenge of your world? What's a manageable project you could launch today that starts you on the path toward achieving that longer-term goal?

Sleep for Success

For the past few decades, people in Director roles at IDEO have known that brainstorms hum in the morning, when energy and creativity seem to peak. Only recently has the idea occurred to us that with a good nap in the middle of the day, perhaps you could get *two* peaks, like having two mornings in the same day. We haven't done any scientific experiments yet, but it seems worth looking into. Certainly, some of the most prolific and creative people in history have been daytime nappers.

Luminaries from Thomas Edison to Winston Churchill have sworn by the rejuvenating powers of a good nap. Churchill began his day with a five-hour morning shift. Smack in the middle came the hearty lunch and ample nap, followed by an evening of work and meetings that often ran till 2 A.M. Albert Einstein said that naps "refreshed the mind" and helped make him more creative. Brahms napped at the piano and da Vinci between brushstrokes.

> With a good nap in the middle of the day, perhaps you could get *two* peaks, like having two mornings in the same day.

Today, while the vast majority of corporations consider napping unconventional at best, science has come down firmly on the side of the power nappers. NASA believes in naps, and a number of recent studies have shown that naps can have a tremendous restorative effect. One Harvard study showed that a midday nap reduced "information overload." Naps appear to improve the brain's ability to learn a variety of tasks and consolidate memories.

I'm a firm believer in power napping, and I feel fortunate to have a metabolism that allows me to plop down and take a twenty-minute nap whenever I need one. By chance, it turns out that twenty minutes may be the magic number, at least in the business environment. That seems to be an optimum amount of napping time to awake refreshed without feeling groggy.

There's a Zen quality to being open to the power of a good nap or sleep. Sometimes the best thing you can do for your performance at a critical meeting or presentation is to sleep on it. Union Pacific calls napping "Fatigue Management." Naps used to be strictly forbidden at the railroad. Now they're seen as part of a proactive, innovative policy to reduce accidents.

Several years ago, Craig Yarde, president of Yarde Metals in Bristol, Connecticut, noticed some of the workers at his 330-person company taking catnaps in their cars or at their desks. Instead of firing them— or even firing off an angry e-mail—Yarde polled the employees and asked them if they'd like a comfortable place to snooze. The employees said yes (of course), and now they have a "nap room" with semi-private recliners they can use anytime.

We're not quite that far along at IDEO, but when we redid our corporate lounge not long ago, my brother introduced the space to our staff as a place to "play video games, watch TV—or take a nap." Sure enough, practically the very next day I walked by an IDEOer taking a snooze.

As I write this, entrepreneurs are trying to capitalize on the need for napping, with daytime "sleeping salons" in New York City. Napping is not a fad. Biologically, there's a letdown after lunch and into the early afternoon. You can blunt it with caffeine, but sometimes there's nothing better than a ten- or twenty-minute nap.

Still think napping during working hours weird? Nikola Tesla reportedly slept as little as two hours a night, making up for it with frequent daytime naps. Thomas Edison slept about five hours a night, plus naps. Margaret Thatcher, John F. Kennedy, and Buckminster Fuller were all power nappers.

If you let go of prejudices or preconceived notions, clearly you can imagine napping as a powerful tool in the hands of an ambitious and creative person. I don't have any pat solutions to napping in a corpo-

JFK, Buckminster Fuller, and Thomas Edison were all power nappers.

rate environment, but I do believe, when it comes to creative work, that a quick nap can help put you in the right frame of mind. So whatever you do, remember that along with assembling a talented team, Directors need to consider another component to a successful project. Find a way to make sure your team gets a chance to recharge their batteries. The cost to your company will be negligible and the upside may be more than just increased productivity. I know more than a few writers and creative types who report they nap when they get stuck, then wake up and—voilà!—they've got the solution.

Take my advice: Sleep on it.

The Deep Dive for Total Immersion

Years ago when IDEO thought of itself as first and foremost a product-development company, we started doing what we called "Deep Dives." The idea was to get a running start on a project, to immerse ourselves in observations, brainstorms, and prototyping—to accelerate the innovation process. During a Deep Dive, the team temporarily sets all other issues aside in order to launch an intense, exhilarating exploration of one specific challenge for just a few days. Perhaps our most famous Deep Dive was the challenge laid out to us by Ted Koppel's crew at ABC News and watched by some 10 million viewers: design a better shopping cart in four days.

Today, our Directors recognize another advantage of Deep Dives. They're still about electricity, about applying energy and enthusiasm and focus to a problem. But as our work has become more complex and shifted toward more collaborative efforts with large companies and organizations, we've noticed another benefit. The intensive process pulls people together. As Maura Shea of our London office puts it, Deep Dives build consensus. Of course, you flesh out good ideas during a Deep Dive, she says. But equally important, the intensive process may help shift mentalities in an organization.

Team members experience new ideas firsthand. They're immersed in the design process—doing firsthand observations and testing the prototypes themselves. Quite simply, the Deep Dive helps drive good ideas within an organization—and get them adopted. It diminishes the need to persuade. The velocity and the depth of collaboration reduces typical "us/them" considerations. Team members see environments and problems in a fresh light. The question becomes *how*, rather than *whether*, to change.

Directors intuitively understand the importance of making these emotional connections. We're human, after all. Exploring new ideas isn't just about making better things and services. It's about changing people's behaviors and attitudes. Which may be the most critical step in innovation.

Familiarity Breeds Honesty

I recently had a chance to meet Robert Benmosche, chairman of Metlife, America's largest life insurer. Benmosche regularly holds "town hall" meetings with groups of Metlife staff members just to open up more lines of communication in that giant corporation. "The better I get to know you," the charismatic chairman told the Metlife Business Acumen group I was among, "the more comfortable you'll be giving me feedback, even criticism." And feed back they did, as the group peppered him with challenging questions. How many senior executives would have the disposition or the self-confidence to voluntarily allocate a big piece of their time giving rank-and-file employees more opportunity to criticize them? In Metlife's case, the increased communication seems to be working. After improving both sales and employee morale during his first few years, Benmosche took the company public in 2000, then doubled its stock price in spite of the bear market at the time. Who would ever dare suggest he is wasting his time? Maybe Benmosche has figured out that the intangible benefit of building relationships with his team shows up pretty tangibly in the long run on the bottom line.

Does your group have an innovation Director or two? A catalyst to keep innovation buzzing year-round? It's everybody's job in the company to be on the side of innovation, but having a few people wholeheartedly adopt the role of the Director can be worthwhile.

Can you be that person in your organization? Or perhaps act as the innovation Director on your project team? The role doesn't have to be about formal authority. If word gets out that you're passionate about innovation, that you are always interested in new ideas, before long you may discover that you are quickly becoming the "go to" person for sparking positive change.

CHAPTER 7
The Experience Architect

> The "value added" for most any company, tiny or enormous, comes from the Quality of Experience provided.
>
> —TOM PETERS

Stroll into the stunning lobby of Vodafone's Lisbon offices, and you can gaze through the windows upon a reflecting pool, where a giant four-meter cube seems to hover just above the water. A screen covering one face of the cube plays a soccer match, while another displays a news program. Hit the right keys on your mobile phone and you can write, draw, or play video games on "The Cube."

If you had a chance to visit the "Personal Skies" installation at New York's Museum of Modern Art a few years ago, you might have glimpsed a new concept for shared human experience in long-distance communication. Visitors entered a futuristic ultra-white room where, by dialing a phone number, they could see overhead the same sky that would be above the person at the other end of the phone call. Sitting under "someone else's skies" can help you empathize and sense what kind of day they're having.

"The Cube" and "Personal Skies" are two of the wilder projects created by some of our talented Experience Architects—people who focus relentlessly on creating remarkable customer experiences. They give a sense of just how far multisensory designs can take you. And the word "architect" here is used in the broadest sense of the term, since the experiences they create can be small or large, and made up of atoms or digital bits.

A good Experience Architect sets the stage for positive encounters with your organization through products, services, digital interactions,

Vodafone's floating cube is a designed experience you control from your mobile phone.

spaces, or events. Experience Architects design not only for customers but also for employees. Their experiences stand out from the crowd. They keep you from being relegated to the commodity world, where price is the only point of comparison. They engage your senses, incorporating tactile sensations, orchestrating the clever use of sound, searching for opportunities to add smell or taste.

And here's the good news: You don't need an industrial-design degree from Pasadena's Art Center or a master's in architecture from Yale to be adept at the Experience Architect role. Near my house, for example, is a bustling, family-run, European-style café (attached to a delightful independent bookstore) that puts all the pieces of a well-designed experience together, while making it look easy. Café Borrone serves up simple, tasty food from the open kitchen. Their efficient young staff circulates through the crowded tables with an air of easy familiarity. The menu—handwritten in colored chalk on rough slate tiles—changes just enough each week so that it stays fresh. The restaurant is sited at a central location that attracts a lively crowd, ranging from Stanford students to septuagenarians. A pleasant cacophony of voices blends with the sound of the espresso machine inside and the fountain outside. A brick plaza makes full use of the California sunshine. Although there's no "main attraction" to the Café Borrone experience that draws your attention, the sum of the parts is somehow both soothing and stimulating at the same time. I doubt that owners Rose and Roy Borrone have a background in design, but they clearly have a knack for playing the Experience Architect. And as a result, there's almost no place I'd rather spend an hour on a Sunday afternoon than at their welcoming café.

Some of IDEO's best experience designs were created for one-time client events. I especially loved one we did for Snap-on Tools. Drama doesn't normally come to mind when you think of mechanics' tools, but Paul Bennett, Owen Rogers, and the rest of the project team orchestrated a celebration of design directions for the toolmaker that bordered on installation art. We rented an abandoned auto-repair shop for the evening and surrounded it with classic cars from IDEO's personal collections. The experience started in the "reception" area (where the auto repair bays had once been), which had been painted black and then filled with large, colorful, point-of-view statements that set the tone for a journey through the rest of the space. Then, in a choreographed tour from room to room, our client team witnessed a blend of prototypes perched on oil drums, intermixed with first-person video clips from mechanics talking about their love of tools, all combined into a compelling and interactive performance that got the company—and us—excited about their future.

Paul Bennett is one of our most gifted Experience Architects, and he frequently reminds us that the strongest experiences have a deep-seated authenticity. One essential role of the Experience Architect is to be the host who never forgets that giving something special to customers is both good business and good karma. Experience Architects view the world as a stage. They believe in the movable feast, in bringing services or products "nearer" to their customers. They see services and even products as journeys to be mapped. Finally, they have a talent for finding the experience in everything, even what might otherwise seem to be the most run-of-the-mill products.

Experience Architects fend off the ordinary wherever they find it.

When you're in the "zone" of being an Experience Architect, you view the world through a simple lens, searching out experience elements in the status quo that are negative—or merely neutral—and then looking for opportunities to fine-tune them. One way to get started in this role is to look at every aspect of your business and ask, "Is this ordinary, or at least slightly extraordinary?" Experience Architects fend off the ordinary wherever they find it, fighting against the forces of entropy and commoditization when it comes to their team or their organization. Asking this question is a remarkably simple and effective approach. How's the experience of calling your customer-service line: ordinary or extraordinary? What's the experience of a first-time customer? You can also apply this methodology to the experience inside your company. How tasty is the menu when your hot project team has a noontime meeting: ordinary or extraordinary? How extraordinary is the first day as a new employee? Having spotted such opportunity areas, Experience Architects then figure out how they might turn the ordinary into something distinctive—even delightful—every chance they get. Sound expensive? It doesn't have to be. On the contrary, it can be remarkably profitable. Ask FedEx, or Callaway Golf, or JetBlue Airlines, or American Girl, or any of a hundred other companies who shook loose of the ordinary and realized extraordinary returns.

Little Experiences

We've worked with universities to inspire the classroom experience, and with hospitals to take some of the pains out of the birthing experience. We've strived to improve the customer experience for hundreds of new products and services—from shopping for lingerie to opening a new online bank account. Many of these are highly complex projects resulting in full-bore multidimensional experiences, engaging most or all of your senses. Think about travel and leisure, for example, where some of the most memorable, immersive experiences—from a day at a relaxing spa to an evening at a hot nightclub—not only incorporate visual sensations but also make optimum use of sound, incorporate tactile sensations, and blend in the right tastes and smells.

But designed experiences need not be complex or expensive. To change a customer's experience with your service or product, start by taking little steps. Changing a single ingredient can sometimes make a big difference. Take antifreeze, for example, the handy liquid that keeps your radiator from rusting and your engine from freezing up on cold winter mornings. The antifreeze in your car's radiator is typically a pure commodity—an industrial chemical like ethylene glycol. In the San Francisco Bay Area, where I live, this commodity-in-a-plastic-jug costs about $7 a gallon at the local auto parts store. As far as I can tell, the only two things that distinguish one offering from another are the color of the plastic bottle and the quality of the packaging graphics. Of course, your car doesn't notice either one.

So when the antifreeze in your radiator gets, say, a quart low, how do you top it up? First, any mechanic can tell you that, for optimum performance, you don't want a radiator filled with 100 percent antifreeze. Your car comes from the factory with a mix of 50 percent antifreeze and 50 percent water, and you want to preserve that ratio. So you're supposed to take the cap off the radiator and peer down into it (or, on most newer cars, the overflow reservoir), estimate how low it is, divide that amount by two (to maintain the 50/50 mix), and add that much antifreeze. So now you've added roughly a pint of ethylene glycol (without the aid of any measuring device) and need to add an equal amount of water. For most people, here's where the garden hose comes into play. And how do you know when the radiator is full? Right about the time a 50/50 mix of water and ethylene glycol splashes up into your eye.

Not the greatest customer experience, right? But a few years ago, one enterprising company figured out how to build a little better experience into their commodity. How? With *premixed* antifreeze. No estimating. No mixing. No chemicals splashing up into your eye. And here is the best part for Experience Architects everywhere: The premixed stuff, even though it is openly watered down, *still sells for $7*. In other words, by turning their commodity into an easier-to-use product, they've found a way to charge you $3.50 for half a gallon of water. Customers are happier because they don't have to premix the nasty chemical. Margins are great because water costs practically nothing. It's a win-win solution because the makers of this new, improved antifreeze solved a latent customer need by removing one cumbersome step in refilling your radiator. Sometimes, by saving your customer even one extra step, they'll reward you for giving them a better experience.

Do parts of your business resemble commodities? Could you turn something you produce or provide into a better experience for your customers? It pays to seek out new avenues to improve the experience. Make your offering ever-so-slightly distinct and it will cease to be a commodity. And who knows, you just might be able to charge more for less.

Trigger Points

Wise Experience Architects know how to focus their energy. If you set out to make *everything* better about your product or service, you may end up with a gold-plated offering that few customers can afford, or with unfocused features few will fully appreciate. So start by asking what's truly important to your customer. The answer may be something small, irrational, elusive, and completely surprising. But finding that answer is often crucial to your success. It's often just one or two essential elements. We call them *trigger points*.

For example, it took the premium hotel chains most of the last century to figure out what, in retrospect, is hilariously obvious: It's the bed, stupid! Don't get me wrong—I love a great hotel with lots of amenities when traveling with my family or luxuriating at a resort. But when I'm checking into a Cincinnati hotel late in the evening, and checking out at 7 A.M., *it's all about the bed.* Many of the major chains have historically been indistinguishable on this point. There are a lot of nice business-oriented hotels out there, but I have trouble keeping some of them straight when I've stayed five nights in four different hotels during the same week, as I did recently. Westin was somehow the first to realize that the bed was more important than just about anything else on a quick overnighter, and they delivered the goods. They named it the Heavenly Bed: a layer of down beneath, a couple of layers above, plenty of soft pillows, and some quality cotton linens to top it all off. I get a great night's sleep in those beds, and that fact makes me willing to overlook many a shortcoming.

I'm here to report that this one small improvement *does* sometimes make a huge difference. For instance, I did that overnighter in Cincinnati recently, and I noticed that the wallpaper in my room was beginning to peel and the bathroom sink was cracked. But you know what? *I couldn't have cared less.* I sank into that Heavenly Bed and slept like a baby. Of course, the other major hotels have noticed and Marriott, for example, now looks like it may beat Westin at their own game. But all other things equal, give me a choice of business hotels and I will always pick the one with my favorite bed. So search out a key trigger point or two at your company—and then make it noticeably better than the competition.

Now, the alarm clock on the bedside table—that's another trigger point. I'm willing to bet that the prized overnight business traveler cares a lot more about the alarm clock than more expensive amenities like a second showerhead in the bath. And here's where nearly every hotel in America seems to have missed the boat (never mind the $700-a-night hotel room in Paris that had no clock of any kind). Maybe it's not the hoteliers' fault. It's the clock radio people who have thrown in so many superfluous features that the alarms are now impenetrably difficult to figure out the first time you go to set them. And unfortunately, most hotel guests are first-time users of whatever clock radio they encounter in tonight's room. After about five minutes of wrestling with a user-unfriendly alarm clock, I simply give up trying. (I carry two backups in my suitcase for this worst-case scenario.) Ever wonder why the quaint custom of hotel wake-up calls still exists? It's not customers clinging to tradition. It's those darned alarm clocks.

Ever wonder why the quaint custom of hotel wake-up calls still exists? It's not customers clinging to tradition. It's those darned alarm clocks.

Nor am I alone in my frustration and disappointment. The folks at Hampton Inn recently examined more than 150 models in their search for a simple, functional alarm clock for their hotel rooms. When none fit the bill, they took the extraordinary step of designing their own. I consider it a pretty significant failure in the consumer electronics industry that a hotel chain is forced to create its own alarm clock, but that's exactly what happened. They took out all the useless features (like dual alarms) and

Frustrated with every clock radio on the market, Hampton Inn designed its own.

made it dirt simple to set correctly. Then they labeled the radio presets with a clear and simple graphic, so you can select rock or jazz or classical even if you don't know the most popular local stations. What was the result? Not only are customers happier with the simpler models, but they are asking to buy the clock radios for use at home. I don't know how nice their beds are, but Hampton Inn has certainly hit another trigger point. Give me a reliable alarm clock and I will rest just a little bit easier. Save me from the embarrassment of oversleeping and I will likely return.

When it comes to the customer experience, sometimes you gain an edge with just one or two trigger points. Fixing a problem or designing a great experience around those trigger points can be very rewarding.

Mix It Up

The role an Experience Architect can play in re-creating a retail store can sometimes be broader than it would be with a single, physical product. Even so, one critical element can still transform the experience. Consider the venerable ice cream store. For much of America, it's a business model that seems to hold few opportunities. Sales at industry stalwarts like Baskin-Robbins and Dairy Queen have been relatively flat in recent years. Not so for Cold Stone Creamery of Tempe, Arizona, the leading franchise of mix-in ice cream shops, with more than 1,000 stores nationwide. Cold Stone is growing like gangbusters by celebrating the ice cream *event* rather than the product. The company is very explicit about its edge, claiming it offers "the ultimate ice cream experience."

If you've been to their store even once, you know how it works. You pick out your flavors from the neatly arrayed fresh ice creams in the glass display case, as well as toppings from fruit to candy, nuts, and chocolate syrup. Then your personal ice cream maker wields two polished metal spades, twisting and turning your custom-chosen ingredients on a slab of polished, frozen granite—the cold stone. Voilà, the ice cream mix-in. The process is both entertaining and slightly hypnotic, like watching the saltwater taffy-pulling machine at the beachside boardwalk. Kids (and parents) are invited to experiment in a light-hearted, fun atmosphere, and the company's ice cream spaders have been known to break into song at some locations. With the dozens of ice creams and toppings available, there are literally thousands of possible combinations and permutations to get exactly the ice cream you desire. "This is where you get to invent your ultimate indulgence," says Cold Stone. "The granite stone is your palette and you're the artist."

What I find intriguing about this story is how direct the company's management is about promoting its stores as an experience. The ice cream ingredients are clearly important, but the defining element is

An entertaining ice cream experience helped create both marketing buzz and growth through innovation.

THE EXPERIENCE ARCHITECT | 175

the experience—what the company calls "ice cream innovation and excitement." Which is another way of saying that a good Experience Architect starts with the same raw materials as others, but then mixes in something original and memorable. Cold Stone Creamery isn't perfect—I'd like to see the granite slab turned into more of a stage, like a sushi chef shaping and slicing his creations. But I think it works. Nor is it an idea limited to dessert. I expect to see more franchises and restaurateurs turning tired old routines into fresh, fun events.

What's next for breakfast, for example? Nutritionists tell us it's the most important meal of the day, and maybe it's due for an overhaul. You see companies out there trying to lure us with breakfast burritos, breakfast sandwiches, even breakfast cookies, but I think there's still plenty of room for innovation. And if it's the most important meal of the day, why are Americans only willing to devote something like six minutes to it—and even then, only while multitasking? For example, I can make a bowl of cereal at home in—no kidding—less than sixty seconds, and yet I am told by food-industry experts that cereal is "not convenient enough." It's not very portable, for one thing, and very difficult to eat while walking or driving. And apparently, that matters. But if people are willing to wait in a long line for their morning coffee or for a fun ice cream experience, then I think there's an untapped opportunity for a different breakfast experience. So if you're in the food industry, put on your Experience Architect hat and start thinking about breakfast. I believe the rewards are out there for someone who creates a compelling experience and then markets it effectively.

Packaging Experiences

Many products and services plateau because the industry gets lulled into complacency with the status quo. They become ripe for innovation, waiting for someone to break through the logjam. What they need is a good Experience Architect to see the way to fashion something new. Remember: It's not that your customers don't notice the flaws. It's just that they figure that's the way it is.

Take paint cans. For decades, everyone has known that paint cans are lousy. Cumbersome to open, difficult to grasp, and ridiculously prone to spill, they don't seal easily or well. Then the lids rust, adding a little extra brown pigment to your Navajo White wall paint. Frankly, paint cans just don't work. But for countless years, the paint can, based on the 1810 invention of the tin can, remained stuck in some kind of nineteenth-century time warp.

Then, suddenly in the last few years, two companies started probing the experience of painting. First Dutch Boy took the initiative of redesigning the paint can. Their new plastic paint containers are cube-shaped, making them easier to ship and store. They're also easy to hold, open, and pour. There's a meaty handle (unlike the wire on most paint cans that cuts into your palm). The top screws on and off, making it easy to open and seal (yes, customers do store half-empty paint containers). Finally, the interior spout makes for better pouring and less dripping. Nearly every one of these innovations improves the painting experience. Was this rocket science? No. But it did take initiative and good design to come up with a better paint can. And a surprising number of decades.

Innovations often hit in flurries, especially after a long hiatus. When one vendor introduces a genuine improvement, competitors are awakened and forced to respond, often innovating different areas of the experience. That's what I like about Benjamin Moore's painting innovation. They dove into another broken part of house painting we're all familiar with. Bringing home paint chips doesn't really tell you whether the color is right. If you're like me and want to compare several different color variations, you need to buy a bunch of quarts—both an expense and a waste. We just went through this process for repainting the exterior of our house, and it took us *seven* quarts until my wife and I were sure we liked the first one (no kidding). And you can't just toss those six unwanted quarts in the trash, either—they're an environmental hazard, so you have to cart them off to the recycling center.

Then Benjamin Moore came up with an amazingly simple but useful innovation. They made it easier and cheaper to prototype colors. They call it "Two Ounces. Too Simple." Two-ounce color samples in clear plastic see-through jars. They're ideal for a small disposable brush.

TURN IT INSIDE OUT

Got a tired, old experience that needs redesigning? Don't have the foggiest idea where to start?

To shake up your thinking, try something radical. Maybe you can turn a part of your experience inside out. That's the thought that came to me when these photos of Monica Bonvicini's "Don't Miss a Sec" performance-art toilet first circulated at IDEO. Yes, it's a toilet enclosure constructed entirely with two-way mirrors. While the person inside the glass box might feel like they're on public display, people walking down the street see only a dark, reflective mirror. It reminds me of a dressing room with a dual personality we designed for Prada's flagship New York store. A woman would walk into what seemed to be a transparent glass changing room. Hit a button on the floor and electricity would charge liquid crystals in the glass, suddenly turning the glass opaque. In perfect privacy, the woman could slip into a new dress. Then, with a dramatic flair, she could hit the button on the floor, and voilà, she's modeling the latest for her friends outside.

Changing the behaviors of small parts of experiences can make a big difference. Is there one element in your service that could be turned inside out?

This unique public toilet blurs the border between privacy and exhibitionism.

The two ounces covers a two-foot-by-two-foot section of wall, making it economical and easy to experiment with several bands of color.

In other words, the firm introduced a novel method to prototype new paint colors that is less expensive, more practical, and more environmentally sound. And because of the low-cost, low-hassle nature of the product, I bet people buy several in one trip to the paint store instead of making seven trips to the store, as I did. And once they find two ounces of *exactly* the right shade of Benjamin Moore paint, they're almost certain to buy gallons of that brand.

What I really like about the Benjamin Moore story is that the company introduced the little two-ounce bottles as an experience. You can see it in their stores and advertisements. The little snack-sized clear plastic containers themselves are just part of the overall equation. It's their effect on the whole customer journey—the way they take the headache out of finding just the right color—that transforms the consumer's experience.

There are some valuable lessons here. First, products or services may pass decades in the doldrums before a burst of innovation sweeps in. More surprisingly, when a competitor innovates, other companies aren't necessarily left at the gate. Sometimes there are multiple "experience points" that respond to innovation. Take the classic wine bottle and cork. In the mid-1990s, we consulted with a major wine-and-spirits firm on the future of bottles and corks and came up with practical yet attractive nonglass containers and noncork closures. We also arrived at an unexpected conclusion: The most effective and elegant way to close a bottle would be a beautifully crafted screw top or bottle cap, technology that was once limited largely to rotgut wine and beer.

It wasn't simply a question of aesthetics. The industry needed to innovate. Part of the wine from each year's grape harvest becomes "corked," ruined by failure of the cork. Though uncorking the bottle is a beloved wine ritual and provides a raison d'être for many a Swiss Army knife, it can also be an embarrassment and a hassle to many wine lovers, especially when the cork breaks off in the neck of the bottle. When we first explored this territory, the conventional wisdom was that nontraditional closures and containers were far off in the future.

Guess what? Lots of vintners and wine-supply firms have been innovating like mad. Elegantly embossed screw-off caps are becoming popular for midmarket wines, and we are near the tipping point of widespread consumer acceptance. Meanwhile, wine in a box, once solely reserved for the unenlightened, has started to emerge as a compelling "new" wine container. Ironically, the box has long enjoyed some advantages over the bottle for everyday wines. The wine is actually sealed in a plastic bladder that shrinks as you pour the wine, greatly reducing the oxidation that quickly spoils an opened bottle of wine. Cylindrical single-liter or rectangular three-liter boxes of white wine store easily in the refrigerator and make for easy pouring for days or even weeks after its first use. The boxes save vintners money in bottling and shipping and offer greater space for color graphics or text. Boxed wines consume less space, don't break, and are getting good reviews for value and quality.

Consider the rich history of the traditional wine bottle. It came into use in the early eighteenth century but underwent little change in nearly three hundred years. Now we're in a flurry of small innovations that may play out gradually over the next decade. Look around. What else hasn't changed in a long time? What might make a better or different experience? Opportunities for innovation often lie dormant, waiting for someone to unearth what's always been there. Experience Architects have the patience to see what others have overlooked, and the initiative and drive to come up with new experiences.

Designer Water

Want proof of the value of a savvy Experience Architect? Look no further than the plethora of vitamin-enhanced bottled waters ringing up sales at your local market. Once, drinking water came primarily from the tap. When bottled waters first emerged in the U.S. market, companies emphasized their pedigree or their ancient source. More recently, bottled water of all types has become one of the fastest-growing beverage categories. And now boutique brands have added hip colors, vitamins, funky names, quirky flavors, and even tongue-in-cheek humor to the experience of bottled water. Facing a tough afternoon at the office? Grab a bottle of Glaceau's peach-colored Perseverance. Long night out? You just might want to start your day with Rescue, a green tea blend. Critics call it "designer water," proclaiming style over substance. But it seems pretty healthy, and the half-serious instructions add to the overall effect. The kiwi-strawberry Focus drink suggests that you "apply orally to dazed and confused brain cells," and the cranberry-grapefruit Balance version is "recommended for gymnasts, ballerinas . . . or individuals simply requiring equilibrium."

Sure it's silly, but isn't that the perfect antidote when you're feeling parched and stumbling through the aisles trying to pick a bottle from a sea of choices? Nutritionists may debate the value of injecting water with vitamins and flavor, but the bigger point is that beverage companies have managed to sell a mood and sensibility wrapped in design, storytelling, and humor. Healthy sales statistics—and dozens of imitators—prove it has market appeal. And if you can turn *water* into a designed experience, you can do that with anything. Don't tell me your company has a dull product or service. It's only as dull as your imagination.

Mapping the Customer Journey

One of the things we often do at IDEO when collaborating with client companies on a new concept is to help them map out their customer's journey. We've done it for trains and planes, where the framework of a journey seems perfectly logical. But we've also created journeys for some very un-journeylike subjects. We've worked on finding a "journey" in services or experiences as varied as checking into a hotel, opening a bank account, navigating a Web site, filing a patent, ordering in a restaurant, and making dinner.

One thing we've discovered: The journey nearly always has more steps than people first imagine. For example, the journey of the car-buying experience often has lots of anxiety-inducing steps, with plenty of opportunities to stray off the desired path. Another surprise: The journey often begins earlier and ends later than people realize. We find it valuable to consider the emotional underpinnings that precede a journey. The first step in the car-buying journey is not visiting the showroom or scanning the newspaper ads. There are usually some preceding steps, catalysts that may be tied to a triggering event (a blown head gasket on your current vehicle), a life transition (your daughter is going away to college), or new social pressures (your best friend got a shiny new car and yours is suddenly looking a little shabby). Similarly, a car salesman may be tempted to think the journey is over when the customer drives the car off the lot. But the path to building customer loyalty—and earning customer referrals—involves following through to understand that customer's experience with service, maintenance, and eventually resale. Understanding not only the steps in the journey but also the customer's state of mind at each point along the way can be valuable to any aspiring Experience Architect.

Journeys That Matter

Some journeys can even help save lives.

Take the story of David Kravitz. David was inspired to start his company, Organ Recovery Systems, when his father received a lifesaving organ transplant. As he researched the field, he learned that the path from donor to recipient was rocky at best and woefully lacking in technology. Most transplanted organs were packed in ice in the same Styrofoam cooler you fill with beer for the beach. Meanwhile, people were dying while still waiting for these precious organs. At any one time, for example, 55,000 Americans are waiting for a kidney transplant. Sadly, some of them will never receive the lifesaving donor organ.

So what did Organ Recovery Systems do? Working with IDEO, they looked at the kidney's journey from donation to transplantation. Kidneys on ice don't last. Eighteen hours is about the upper limit. Instead of viewing the act of transport as simply a period of time during which doctors strive to maintain the organ, the team strove for something new. They prototyped and developed a process that more actively preserves the health of the kidney. The result? The LifePort, which gently perfuses the kidney with a cold liquid tissue-nourishing solution. The device also monitors and evaluates the kidney during the trip. Incredibly, the perfusion process nearly doubles the potential storage-and-transit time. Healthier, easier-to-transport kidneys mean that more organs can be successfully transported and more people are likely to receive transplants. The odds are higher of getting the organ to the right place before time runs out. There's simply more time to match up available organs with surgeons and patients.

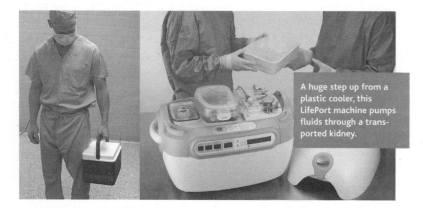

A huge step up from a plastic cooler, this LifePort machine pumps fluids through a transported kidney.

Devices like the LifePort could make a big dent in the huge backlog of individuals on organ-transplant waiting lists and save thousands of lives. Meanwhile, the health care industry potentially stands to save a billion dollars in reduced emergency costs as the number of viable organs increases and transport pressures lessen. All because David Kravitz applied some insight and rigor to improving the journey of a kidney from donor to recipient.

We're proud of our work with Kravitz and hope to collaborate with his firm on LifePorts for other organs someday, like hearts or livers. Journeys do matter, whether it is the customer's journey through your product or service or the journey that brings your offering to the attention of a potential customer. How might you come up with a journey that makes a difference in your marketplace?

Moving Journeys

Experience Architects are acutely aware that the task of designing experiences is constantly evolving, influenced not only by the spread of new technology but also by shifting human needs. The days when most customer journeys began at a fixed location are fading. Sometimes you have to bring your designed experience directly to customers. Even right to the office parking lot.

A while back, IDEO's Smart Space group (designers who create experiences on an architectural scale) teamed up with an ambitious northern California company that had a radical idea for the future of dentistry: bring the dentist to the patient, instead of vice versa. The business plan called for a fully equipped mobile dental van to visit major corporations with thousands of employees. Companies could use it as an extra service for their workers—one that reduced downtime yet cost the company almost nothing. I anticipate that other sophisticated professional services will be plugging into corner spots of corporate parking lots in the years ahead. Time pressures, wireless technology, and advances in miniaturization will help drive this trend. Yet popping wheels on a service or experience is only part of the equation. A successful mobile service can challenge industry traditions. For example, if the dentist's office is suddenly a two-minute walk from your desk, do they need a waiting room? Proximity opens up the chance for the receptionist to call you a few minutes before the previous session is finished, reducing your wait time to nearly zero.

Mobility works, especially with companies that never really had much of a physical presence. For instance, Progressive Insurance, a maverick national auto insurer out of Mayfield Village, Ohio, tackled the major challenge facing an insurance company by figuring out how to settle claims quickly and efficiently. They do it with mobility. Progressive prides itself on Immediate Response, a fleet of Ford Explorers driven by claims adjusters that circle major cities round the clock to provide astonishingly quick service and claims. Progressive's greatest breakthrough is the tighter link it has forged with its customers. Mobile claims adjusters wielding laptops download claim information and do sophisticated on-the-spot analysis to work up estimates. Progressive is also smart about simple stuff. Customers, for instance, are given Progressive cards that break in half to ease the sharing of information after an accident. In other words, they've thought through the experience of what clients must do after an accident. In a business where claims used to be settled in weeks or months, Progressive strives to settle claims in days and sometimes hours. I recently had my first car accident in thirty years, and the idea of on-the-spot service sounds pretty compelling compared to what I experienced.

The lesson is that the Experience Architect can help transform even the most tradition-bound industry. Most giant insurance firms are by nature risk averse, and they've been reluctant to let go of their paper empires and deskbound agents. Progressive's momentum has come from something as straightforward as putting wheels and technology on the experience of filing an insurance claim. Too big a change, you say? Mobility won't work in your industry? Consider the experience "appetizer" Lexus serves up to new-car buyers. When I bought a car from them a couple of years ago, the dealer came to my office for my first scheduled maintenance service. All I had to do was leave my keys with the receptionist and a note hinting where the car was parked. Who wouldn't want that? All the service. None of the hassle. My customer journey for a 1,000-mile maintenance check consisted of the pleasant experience of not having to go anywhere at all.

> Too much time standing in one place is bad for your corporate health. Drive, run, or walk to your customers with technology, information, and personalized service.

My advice? Too much time standing in one place is bad for your corporate health. Drive, run, or walk to your customers with technology, information, and personalized service. Wait too long to take the first step and somebody else will get there first.

The Non-Journey

The metaphor of a journey can be a wonderfully evocative vehicle for capturing the broader, underlying emotional responses that spark a new product or service. That said, Experience Architects don't believe in a one-size-fits-all approach. The nature of this role is to design experiences that fit the unique demands of each new product or service. So even IDEO, which strongly believes in the power of the journey, has learned that no one model fits every situation.

For example, a few years ago, a well-known retailer asked us to explore the shopping experience for kids in their stores. They were contemplating a major redesign, and we collectively agreed that mapping the kids' shopping journey might provide jumping-off points for new store concepts.

We started by spending a lot of time in the company's stores, often down on one knee so we could get the kid's-eye view. What did we notice? The kids' journey wasn't anything like what we'd studied on other projects. To begin with, kids didn't really view the store visit as a journey at all. They couldn't recall what was first or last. They could certainly tell you what was cool. Or what they considered "boring." But they did not perceive the shopping experience as sequential steps, and our hunch was that they would not respond to a store designed as a journey.

Fortunately, we'd hit upon another way of approaching the problem, a different way of telling the story. From our observations, we'd learned a lot about some discrete cognitive states that kids go through in a store, and we built a new conceptual framework around those mental states. Our discoveries took us to a pretty surprising conclusion: that the costly plan to overhaul hundreds of stores might be an unnecessary expense. So instead of one grand interior plan, we gave the client nearly two dozen concise insights about kids to help inform the details of their store experience. Celebrate the presentation of the products, we suggested, the packaging, the merchandizing, and the fixtures. Work on the tables, displays, lights, graphics, and so on.

What was most remarkable about this project was how our belief in

THE HAMMOCK FACTOR

Powerful icons or talismans can help symbolize the state of mind you are trying to convey in designing a new experience. I'm referring not to religious artifacts but to cultural symbols that evoke strong, positive associations.

Just the *sight* of a hammock makes me relax. What similarly powerful associations can you draw upon?

Think about the humble hammock, preferably stretched between two palm trees. It's hard to see a hammock without thinking about relaxation. Similarly, it's almost impossible to sustain a state of stress or worry while lying in one. Having recently retired our kids' play structure in the backyard, I am thinking of putting a hammock in its place as a form of family stress therapy. What's the hammock in your business? What generates such an overwhelming response that your customers can't help but be intrigued?

staying true to observations informed our course. Had we stuck to the familiar "Plan A," we could have settled for a detailed kids' journey and presented our findings in a more typical manner. But along the way we found that, however robust the metaphor of the journey, it simply wasn't the strongest vehicle to mine new ideas for a kids' store. As Stephen Covey likes to say, "The map is not the territory." So even though we love a customer journey map in most cases, we recognized that this was one of those times when you need to set the map aside.

Think of it as sailing off with a compass as your guide. When the stars in the night sky suggest that the compass is broken, the experienced navigator forges a path with the best tools he can find. That, to me, is the strongest methodology: the independence of mind to draw your map and tell your story from the conditions you encounter, not the preconceptions we often bring to a project. The confidence and the humility to change course when new insights tell you that's the right way to go.

Authenticity

Experience Architects have a nose for what's real. They strive for the authentic, individual impressions over the "official" expert. Take restaurant reviewing, for example. Not too long ago, we made our choices primarily on the basis of single make-or-break reviews by local newspapers or the gilded opinions of national guides like Frommer's, Fodor's, or Michelin. The experts told us what was what. Who were we to argue?

Well, it turns out that the reviews weren't always as authentic as one might think. Restaurants often could spot a reviewer and artificially enhance the visit with extra-special food or service. Reviewers themselves were often far from objective. Besides, was it fair to write a review based on one or two visits? Enter Zagat, the populist arbiter of reviews for restaurants and hotels. Zagat has turned reviewing on its head by firing the professional reviewer and welcoming you and me and thousands of others like us. Zagat is a massive survey based on hard-to-fake authentic diner impressions. Zagat surveys pool a phenomenal quarter of a million volunteer reviewers. Its husband-and-wife founders, Tim and Nina Zagat, figured out long ago that authenticity has its own rewards. No, they don't employ any celebrity critics. Instead, they hire knowledgeable editors to design the surveys and decide which restaurants to include, composing "reviews" on each restaurant with cleverly strung-together phrases originally written by the volunteers. It's akin to the People's Choice award. Far from the snobbery implicit in an elite newspaper review, a Zagat mention has the appeal of grassroots reality. Zagat reviews generate buzz in a way seldom found in a "professional" review from the likes of Frommer's and Fodor's. Zagat surveys have become so popular that they have recently expanded beyond hotels and restaurants to include movies, music, and even golf courses.

Real has a genuine appeal. Reality shows may be edited, but their popularity has been largely based on the premise that they are unscripted. Similarly, the questionable but extremely popular HOTorNOT Web site's visceral attraction derives from the simple premise of exhibitionists (or masochists) volunteering a picture of themselves to be voted on by thousands of total strangers.

Zagat, however, was the trendsetter, first minting its "reality" approach in 1979. The company has gone beyond Amazon.com's reader-inspired reviews in part because Zagat vets its reviews with an editor, while many online services have yet to figure out how to weed out "friends-of-the-writer" reviews. Zagat customer reviews generate trust because they are mediated with knowledgeable, professional editing.

Through a deep sense of authenticity, the Zagats have built revenues and a broadening reputation. With the strength of their "real world" brand, they have opened the door to potential new markets. What's next? Reviewing cruise ships? Museums? Auto-repair shops? Dentists, doctors, general contractors? I won't try to predict what markets they'll pursue, but I do know they have untapped value in a company synonymous with "real." Are there chances for your organization to "get real," to tap into the intrinsic rewards and value of authenticity? Think Zagat and tap into the power of *real*.

Just as you can't fake authenticity, you can't fake personality. We all know friends who have genuinely interesting and entertaining personalities. To begin with, they're real, honest-to-goodness individuals.

There is something unique and engaging about how they act, talk, and look. The same is true of the best companies. They summon a mood or spirit. Virgin is irreverent and fun. BMW takes driving seriously. Apple is about iconic design. Ritz-Carlton emphasizes over-the-top service. IKEA offers affordable style. You know them when you see them. Some companies have personalities that make their colors shine, making competitors look a little gray by comparison.

Merit Badging

A dedicated Experience Architect draws on all available sources, especially from their own life experiences. For example, my son was going through some old boxes recently, and dug out the olive-green sash of Boy Scout merit badges from my days as an Eagle Scout. While he peppered me with questions about the meaning of twenty-one different badges like "pioneering" and "firemanship," I was flooded with memories about the great times I had in scouting. Most people don't vie for merit badges anymore, but the trend watchers at Iconoculture say there's an emerging lifestyle pattern called "merit badging." The idea is that large groups of people, having climbed psychologist Abraham Maslow's hierarchy to the point where they feel they have amassed enough *stuff*, are now collecting *experiences*. The idea struck me as both powerful and familiar the moment I heard it.

My brother David, for example, pointed out almost twenty years ago that most of his friends didn't need any more stuff in their lives. David loves to give gifts, and had been filling his pickup truck with presents every year at Christmas, but gave it all up one year and decided to give experiences instead. I don't even remember what elaborate gifts David gave me in the old days, but I do remember every single experience he's shared with his friends for the last two decades. He gathers up a couple of dozen friends (and, thankfully, a few family members) each year and takes us on a small adventure, letting us share the experience with him. We've been to Rolling Stones concerts and monster-truck shows. We've had cooking lessons from a fabulous French chef and learned magic from a sleight-of-hand artist. We've raced go-karts on a giant indoor track and cheered on the San Jose Sharks from a sky-box. Usually, he charters a bus and serves us a catered meal himself on board, acting as a host and flight attendant. Are David's experience gifts a splurge? Absolutely. But that's not the point, since he could choose a picnic on the beach and we'd have just as much fun. David has reached a higher plane in the pantheon of gift giving. Rather than filling our closets with objects we may or may not need, he gives us experiences to share that we can remember forever.

I see "merit badging" all around us. Fifteen years ago, my friend Stuart Graham told me that one of his life's goals is to visit a hundred countries. He is already at ninety-three, with half a lifetime still to go. And I don't know if IDEO Transformation team member Hilary Hoeber has the same ambition, but she did mention during her interview process that she'd visited twenty-five countries by the age of twenty-five. But merit badging isn't limited to adults, either. At age eight, my son decided to read all of the classic Hardy Boys books—in numerical order. He plowed through all fifty-eight of them without reading another book for the better part of that year. My daughter is "collecting" U.S. states right now, so we're trying to take family vacations to states she can add to her list.

Many Gen Xers and Millennials are less interested in accumulating material possessions than their parents were. As new generations redefine affluence, it may be less about what you own and more about what you've done. Races you've run for worthy causes. Marches you've walked in. Events you've volunteered for. The growing popularity of the annual pilgrimage to the Burning Man festival in the Nevada desert is yet more proof of this trend. Burning Man is a truly unique art festival and temporary community that now attracts more than 30,000 people to Nevada's Black Rock Desert for one week every August. There is literally nothing to buy there except ice, tea, and coffee—and even the most impressive art installations get burned to the ground. Yet anyone who has gone to Burning Man will tell you it is a remarkable and extreme experience they will never forget. And things almost perceptibly slow down for a few days at IDEO every August, as an ever-expanding number of IDEOers strive to add the Burning Man experience to their merit-badge sash.

Experience Collectors

What experiences can you offer your customers? Can you encourage them to *collect* your experiences? Visit all your locations? Try each of your offerings? Visitors to Japan may have noticed that virtually every major train station, tourist attraction, and hot springs location has an oversized rubber stamp to imprint on your travel journal or notebook. Stationery stores and gift shops even sell simulated Japanese passports to receive such stamps. Incredibly, the whole country has turned traveling into a game. Like a world traveler collecting stamps on a passport, one can fill a whole book full of these destination stamps without ever leaving the Land of the Rising Sun.

Designer Philippe Starck and hotelier Ian Schrager have teamed up to create eight boutique Morgan's Group hotels around the world so far (and I am sure more are on the way). When I noticed one day that I had already visited six of them, I felt compelled to "collect" the seventh during a recent visit to Miami. Sometime soon, I'm sure I'll find an excuse to complete my collection by staying at number eight, the Mondrian in Los Angeles. Why? Call it the merit-badge factor. Once you've got seven, the eighth seems practically de rigueur. Just like once you've read the first five Harry Potter adventures or the first dozen Tom Clancy thrillers, it's hard to resist the latest installment.

Merit badging as a phenomenon has a life of its own, with or without an actual badge. Classic movie fans may remember a famous scene in *The Treasure of the Sierra Madre* in which Humphrey Bogart confronts a gang of armed bandits, unconvincingly masquerading as federal agents, and asks to see their badges. In the most memorable line of the movie, the bandit leader shouts back, "Badges? *Badges?* We don't need no stinking badges." True, you don't *need* them, but both physical and virtual badges can carry a lot of meaning, symbolic or otherwise. Back when the San Francisco Giants played at frigid Candlestick Park, the stadium was so cold that we would—no kidding—wear ski parkas to a night game in July. It was such an ordeal that the Giants eventually turned it into a badge of honor. Survive an extra-innings night game and you would be awarded with the only slightly tongue-in-cheek "Croix de Candlestick," an icicle-themed orange-and-black metal pin that you could fasten onto your hat or jacket. Believe me—the die-hard fans wore their Croix de Candlestick pins with great pride. A few years ago, the Giants moved to the delightful and warmer SBC Park, but you still see the old-timers in the new stadium wearing hats covered with those badges of honor.

So pick out your best, most loyal customers, and help them become connoisseurs of your product and service. If they have sampled half your offering, tempt them with the other half. Give them a reward, a symbol, a stamp on their passport, a gold star. Your best customers are hungry for new experiences. Help them fill their sash.

Architecting the Extraordinary

An Experience Architect is the right person to remind your organization that the first step in becoming extraordinary is simply to stop being ordinary. To beat the competition, outperform the market, and exceed the norm, you have to create remarkable experiences for your customers, for your partners, for your employees. And if the Experience Architect can help you do that, then word will soon get out that there's something special about your team.

CHAPTER 8
The Set Designer

Every organization (and every employee) performs a bit better or worse because of the planning, design, and management of its physical workspace.

—FRANKLIN BECKER, *OFFICES AT WORK*

The Set Designer persona is hardwired into the collective IDEO psyche. From the beginning, we've had the implicit belief that a creative office is like a well-designed stage or movie set that contributes to the overall performance. And the unofficial Set Designers who constantly tweak the design and layout of our offices know that they're supporting not only our work but also the culture itself.

We've still got the DC-3 wing hanging from the ceiling, handpicked and lovingly polished by an IDEO team several years ago. David Littleton's string of multicolored Christmas lights twinkles above his space year-round. And Roby Stancel still maintains his collection of "weird foods of the world" at his desk in Munich. Sometimes, for kicks and camaraderie, a team will execute a spontaneous makeover of someone's space when they're not around. For example, Dennis Boyle's corner space includes a striped awning from a Parisian café and a fifteen-foot structure that looks like a section of the Eiffel Tower. And one recent weekend when CFO Dave Strong wasn't looking, a team redesigned his space to resemble a local pub, complete with barstools, a dartboard, and a bar-height counter where his desk once stood. I don't know if he plans to keep the pub ambiance permanently, but it's still there so far.

What's the point? We're a company that revels in collaborative, team projects that celebrate the individual. Giving employees more latitude in the shape and character of their workspace helps reinforce a company persona that is fun, welcoming, and stimulating. Does this freedom and creativity make you money? My answer to that question is another question: How many companies really want their offices to be boring, dull, and devoid of energy and emotion?

Inner Space and Sensory Underload

In the original *Star Trek* television series, Captain James T. Kirk, commander of the starship *Enterprise,* introduced the show every week by reminding us that space was "the final frontier." While Captain Kirk spent his life in outer space, Set Designers are dedicated to exploring a different frontier you might call "inner space"—the work and commercial environments where most of us spend the bulk of our waking hours. And why would I consider this kind of space—especially office space—the "final frontier"? Because although a carefully crafted work environment is essential to an innovative organizational culture, far too few companies grasp its importance. Space is among the last things managers think of when trying to revitalize team attitudes and performance.

Sit amid the soaring arches of Santiago Calatrava's amazingly engineered Olympic Stadium in Athens and you wouldn't doubt the value of a Set Designer. Stay in one of the experience-rich, one-of-a-kind André Balazs hotels (like the Standard in L.A. or the Raleigh in Miami Beach) and you'd never think for a moment that the world you've entered hasn't been designed exclusively for your fun and entertainment. But when we walk into most offices, our senses shut down from sensory *underload.*

Bland office environments have become part of the business landscape, something you stop noticing after a while. Business guru Tom Peters rails against "Dilbertville" and what he calls "The Great Blight of Dullness." He says, "Dreariness, from the reception area to the research lab . . . destroys the spirit. It's utterly impossible to imagine people laughing in such settings, or weeping, or frolicking, or producing anything interesting!" We all know exactly what he means, and we know there are better alternatives.

Companies may consider corporate set design the domain of architects, facilities managers, or space planners. Individuals may feel there isn't a lot they can do to influence their space, or that set design is too "soft and fuzzy" for someone of their stature. But if new ideas are your business, if generating energy and enthusiasm matters, then every single person on your team would benefit by adopting a little bit of the Set Designer's outlook.

We all know lousy spaces when we see them, yet all too many companies keep churning them out. You've probably seen those cogs-in-the-wheels-of-industry scenes from old black-and-white movies with seas of identical, factorylike desks and typewriters. Patently bad for creativity and innovation, you might say to yourself. But here in the twenty-first century, all too many of us settle for eight gray hours at a drab cubicle. Oh, we've got computers and mobile phones and networked printers, but the space rarely sings.

Most companies have someone in their Facilities group who oversees all the myriad decisions of where teams and divisions sit and work. Periodically, they may hire interior designers and architects to plan new spaces or modify existing ones. I believe there is room in many companies for more people taking an active role in the nuts and bolts of office space. Set Designers look at every day as a chance to liven up the workplace. They create collaborative spaces for "neighborhood" teams. They gauge how space behaves and make subtle adjustments to keep it responsive to your shifting needs. Set Designers balance private and collaborative space, giving people room to collaborate but

> The Set Designer can be the "X factor" in a company, the intangible element that helps turn around an organization.

also providing the sanctuary of privacy for intensely individual work. They create project spaces, making room for projects to live and breathe for weeks or months. They can help people move and migrate, forming new groups and potent combinations. The Set Designer can be the "X factor" in a company, the intangible element that helps turn around an organization. At heart, we're all Set Designers—and even those microdecisions about how we arrange our own desktop, chair, or even dictionary can make a difference in our daily work.

Some workplaces are so dull that you only have to lift one or more restrictive rules to make an improvement. Perhaps you work in one of those places ripe for innovation. A few years back, I visited a company that employed hundreds of creative people—artists, writers, and graphic designers. When I spoke to management about their work environment, they raised the usual objection: Space is expensive to build and costly to change. But I pointed out one change that could be implemented in less than twenty-four hours absolutely free. For reasons I still can't fathom, they had one of those archaic "nothing should stick up more than four inches above your cubicle" rules. Were they afraid of blocking the "view" of other cubicles? Disrupting the uniform blandness? So I told them to jettison the rule. Blast out an e-mail. And while they were at it, launch a contest for the individual or team with the most colorful and distinct workspace. There was still plenty of room for improvement, but it seemed a good start. And it was hard to beat the price.

Of course, corporations that more fully embrace innovation can do a lot more than just relax a few rules. For example, Procter & Gamble has recently embraced the power of design and innovation, and decided it needed a special place to nurture fresh new innovation initiatives. We helped them design a space they call The Gym, a 10,000-square-foot innovation center. One key decision was to build The Gym in a location near the majority of P&G Cincinnati employees—in other words, they wanted an "off-site" to be built "on-site." Collaboration is key. Three large learning areas called Initiative Spaces feature an open design with easily movable furniture, lots of low-tech surfaces to write on, and stick-up Post-its for shoulder-to-shoulder collaboration. The Gym is a blend of the ultramodern and the casual—there's an informal café and the latest in information and display technology. Interestingly, P&G helped come up with some of the cultural guidelines to help new-to-innovation employees make better use of the space. For example, visitors are encouraged to check their traditional roles and attitudes at the door. P&G sees The Gym as a place where teams can collaborate and ideate, not just about new products and services but also about the process of innovation that keeps it ahead of the curve.

The right space can be a compelling tool in sustaining continuous innovation.

As mentioned in Chapter 6, Mattel also created an "on-campus off-site" space to help spur its Platypus team to innovation success. Even in companies not quite ready to dedicate a space to innovation, however, there are still other reasons to nurture the Set Designer role. For example, every organization worth its salt needs to recruit great people, and space can be a big draw. During the toughest Silicon Valley talent wars of the late 1990s, when the first Internet boom was at its peak, we had several new hires say they joined IDEO because they toured the office and it seemed like a place they would like to work. And if a more compelling set design for your office can raise the quality of people willing to work for you—or make even a small dent in unwanted staff turnover—that's a result even an accountant could love.

Since it's difficult to empirically prove that great space directly increases revenues, sometimes a Set Designer has to push hard to get their company to invest in space. That barrier isn't nearly as high for a Vegas casino. They've been measuring the return on innovation for space improvements since the 1950s. Make hotel guests walk through a zigzag of slot machines on their way to the elevator, instead of a straight line, and the "take" may go up by a measurable 0.7 percent. Add keno displays to the casual restaurant, and a break-even operation suddenly becomes profitable. It's true that outside the exotic field of casino real estate, the value of great space isn't always so measurable. But there's still a noticeable relationship between energized spaces and energized teams. Industry-leading companies like Pixar and eBay understand that their highly collaborative environments are central to the happiness and creativity of their talented staffs. Set Designers recognize that when it comes to companies that depend on freethinking, idea-sprouting individuals, space plays far more than a supportive role. Space can become the place where ideas take shape and opportunities emerge.

Here's an example. A few years ago, the creative arm of the BBC in charge of documentaries and true-life programs came to IDEO with a problem that had been nagging them for some time. The twenty-seven men and women who dream up this division's new programs spent their workdays in an environment they considered second-rate. The staff seemed embarrassed by their drab quarters and felt the space added nothing to the energy of the team. Our London office was asked to collaborate with the BBC to tackle the problem. Since the television industry revolves around projects, they opted to focus on space improvements that could enhance teamwork.

One of the first steps IDEO took was to create a dedicated brainstorming area, with everything from whiteboards and oversized Post-its to robust presentation technology, comfortable seats, and stylish café tables. A place to brainstorm may sound like a small thing, but it can make a huge difference. Within weeks, the brainstorming room had become so popular that teams often used it in the evenings, after other offices had long since gone dark.

A customizable project space was another immediate hit, a room designed to reflect the sensibility of a particular audience. For example, during the time that the team was focused on attracting more teenage viewers, creative members could plunk down in a beanbag chair or cushy sofa to watch the pilot version of a new show, surrounded by posters of pop bands, hip magazines, and other artifacts typical of a teen's bedroom. The idea was that they could gain insights and empathy by watching a pilot or prototype while immersed in an environment similar to

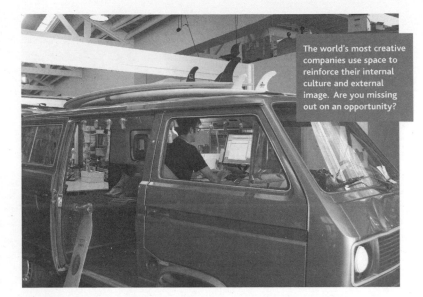

The world's most creative companies use space to reinforce their internal culture and external image. Are you missing out on an opportunity?

the setting that their youthful audience was watching from. While developing a show targeted at the older set, the same project space would be transformed into a homey, familial scene with the mantel clock over the simulated fireplace set to 9 P.M. and a pot of tea on the table. Strangers to the vagaries of television may find the concept a bit wild, but the project room quickly became a favorite spot to critique new ideas and pilots. A "hub" space offered something else that had been missing in the old offices—the chance to mix with colleagues and get ideas from outside one's own group. A cross between a café and a private theater, the hub became a popular nexus for many BBC staffers to grab a cup of tea and watch English football or a new pilot.

Perhaps the biggest breakthrough was the most direct: establishing a room dedicated to pitches for new "programmes." Before their space was reinvented, the team rarely invited people in for meetings, instead renting conference rooms outside the BBC or scheduling the meeting at the more presentable office of an independent producer. Suddenly, they had a fully equipped meeting room with the latest in technology.

A skeptic might say "So what?" Individually, these changes don't sound exceptional. But Set Designers recognize the power of the sum of the parts. Collectively, these project-based and collaborative spaces lent the team new momentum. The BBC reports that this group won significantly more program commissions after their space was redone. So—creatively speaking—sales were way up. And if a creative group can redouble their efforts, who could argue with the cost and effort of the space improvements?

Whatever your business, you've got to continuously bubble up new ideas and ways to make them a reality. Don't forget to make better space part of your formula. That you may not be able to measure every element of what makes you more innovative doesn't mean you shouldn't try something fresh.

Enter a New Arena

Growing up in northeastern Ohio, my "hometown" baseball team was—for better and mostly worse—the Cleveland Indians. Throughout my childhood, all the way through my twenties and thirties, the Indians were one of the sorriest teams in American baseball. The Chicago Cubs admittedly had a worse win-loss record some years, but at least the Cubs were lovable. The Indians were just *bad*. Not only did they never win their division during that interminable stretch, but they rarely broke .500 for the season. When Hollywood was looking for a sad-sack team to lampoon in the comedy *Major League,* the Cleveland Indians were the easy and obvious choice.

But in 1994, the Cleveland Indians went through a remarkable transformation. They had the same coach and essentially the same players that year. They played in the same town and drew from the same pool of fans. But there was one big catalyst for change in '94: the stadium. The Indians finally moved out of the gigantic and drafty old Municipal Stadium on Lake Erie, where they had capacity for 80,000 fans but often played to less than a tenth of that. Abandoning that aging behemoth, the Indians moved into the endearingly compact and airy 40,000-seat Jacobs Field, known locally as "The Jake." It was built right in the heart of downtown—and was filled to capacity on game day for the first five

Sometimes changing the arena can change a team's performance.

years. Suddenly, Cleveland's "Bad News Bears" had, for most of that season, the best record in American baseball. As luck would have it, the 1994 baseball strike canceled that year's playoffs. But in 1995, for the first time in more than forty years, the Cleveland Indians won the pennant, disproving the theory among many Ohioans that the Indians were waiting for hell to freeze over before they'd get to the playoffs.

If the Cleveland Indians' transformation from one of baseball's worst teams to one of the best was catalyzed by changing their stadium—by changing the working environment of their team—then maybe what your team needs, too, is a fresh stage to find its inner greatness. It doesn't matter whether your talent resembles the hapless Cleveland Indians of my childhood or the perennially mighty New York Yankees. If simply redesigning your work environment would make your team more of a winner, who wouldn't make the change?

By the way, Jacobs Field not only helped reinvent the team and earn profits for the owners, it helped reinvent the whole city. The new stadium spurred a wave of renewal that included the Rock and Roll Hall of Fame, a world-class science museum, an IMAX theater, and a series of small improvements that have lifted the spirits and the economy of the whole town. So although the new, improved Cleveland Indians aren't in any pennant races right now, the stadium has had a lasting and positive impact on the team and its community. Smart Set Designers can make a difference.

Innovation on Wheels

Just as a well-designed stadium has its own vitality, a good kindergarten classroom reverberates with a steady beat of originality. Five- and six-year-olds haven't been trained out of their natural creative abilities, so kindergartners are always coming up with new ideas. Most kindergarten classrooms are high-energy, perpetual-motion environments. Everything in the space moves to accommodate or support a desired activity. Desks are shoved out of the way to create new spaces, corners of the classroom are adapted for special projects. Mobility and flexibility contribute to the sense of exuberance, excitement, and creativity.

Kid stuff, you say. Well, I'd venture to suggest that some of America's most creative companies take a measured kindergarten approach to space. Take George Lucas's Industrial Light and Magic. The firm has been a relentless innovator in film and special effects for more than three decades. ILM's out-of-this-world success in the original *Star Wars* helped Lucas build a film empire in northern California. One of the multibillion-dollar firm's secrets is that it is repeatedly willing to take huge technical, creative, and financial risks in pursuit of visual-effects breakthroughs, a strategy that few Hollywood companies are willing to embrace.

Because the creative team members at ILM are kindred spirits to our group at IDEO, we've spent time with them trading stories, ideas, and occasionally staff members. There are a dozen reasons for Lucas's continuing success, but here's one you may not know about: ILM has a tremendously flexible approach to the workspace. They staff by project, with huge fluctuations in workload as new film projects ramp up and drop off abruptly when it's "a wrap."

Not only are the ILM spaces incredibly creative and stimulating, with scale models surrounding you at every turn, but the firm is also prepared to remap their eclectic office environment quickly to suit the changing needs of the project teams. At ILM, the Set Designers are constantly at work. A senior executive there recently told me, "I sometimes think ILM means 'I Love Moving.' We did close to 800 moves here this year, and we only have 1,200 people on the whole campus." If that sounds like a lot of cost or hassle, consider this: Staying in the same place can be much more expensive if it hinders your business's ability to tackle new projects with fresh teams. And having a fluid work environment reduces the chance of people getting "stuck in a rut," following the same thought patterns and talking to the same small circle of people.

Is your workspace too static? Do you need to move people around? Are you letting the natural human resistance to change stop you from launching new, more creative combinations of workers?

Maybe, like the kindergartners, you need to bounce around a bit.

Big Moves

It's the Set Designer's dilemma: What if you literally can't afford to move from an average building or bad neighborhood? What if, for instance, you're stuck in a long-term lease? Sometimes what's needed is to shift the question. To search for new space options even when none seem to exist.

While working with clients at Levi Strauss & Co. early in my career, I first heard how the reinvention of their office space began with a fireside chat. In the mid-1970s, Levi's CEO, Walter Haas, Jr., was regretting his company's recent move to a thirty-four-story high-rise tower in San Francisco's Embarcadero Center. What was a great office space for some companies just didn't fit for Levi's. There was no proper lobby for the Levi's team, since they were mixed in with other companies. Getting a cup of coffee often required an elevator trip. Glass-walled offices brought neither privacy nor camaraderie. But despite the visceral dislike of the new offices, Haas felt trapped. He felt locked into the current deal by the company's twenty-year lease on the high-rise.

One evening in the Montana wilderness, Haas confided to real-estate developer Gerson Bakar that he felt as if he were in an ivory tower and the company had lost its family character. Sitting around the campfire that night, Bakar replied with an extraordinary offer. He proposed finding a campuslike setting near downtown San Francisco, building the new headquarters of Haas's dreams, taking over his Embarcadero lease, and not charging Levi's any more than they were currently paying.

Back home in San Francisco the next week, the developer backed up his initiative with follow-through—and the help of top architects. Several presentations later, Levi's executives and employees signed on. Thus was born Levi's Plaza, an understated campus of low-rise brick buildings at the foot of San Francisco's Telegraph Hill that employees and executives loved. Working closely with Levi's during that period, I had the chance to spend time in both the old and new headquarters. The two settings couldn't have been more different. The old was elite and aloof, the new open and conducive to collaboration. Outdoor terraces on each floor were great for impromptu meetings and coffee breaks. Workers and visitors adored the generous plaza and bountiful three-acre park, with its inviting fountains, grassy hillocks, trees, and flowers. I was especially envious of Levi's Plaza because my company stayed behind in Two Embarcadero after Levi's moved out.

I found their story inspirational because it proved that with the right attitude—and a Set Designer's flexibility—you can topple intractable barriers to changing space. Who says you're doomed to stay in a lousy building? Levi's management was honest enough to admit it had blundered in moving to the Embarcadero tower. They were also willing to consider a radical proposal.

In a sense, the firm's mistaken Embarcadero move was a prototype. The lesson? Sometimes you have to have the imagination and courage to recognize that you need to move on to something better.

Connecting Spaces

Set Designers care about the intersection between space and human behavior. They care about making connections and following through on intentions.

Consider, for instance, a couple of common preconceptions. Hallways are for moving through quickly. Classrooms are for closed-door lectures. Right? These long-held presumptions were put under a microscope recently by an IDEO team on a project for UC Irvine. They didn't have a lot of time. They did observations in hallways and the classrooms. They met with students and professors, and roamed the public and private spaces.

What did they discover? That students looked upon lectures as a meeting point, a jumping-off point for more learning. Class was not just the place where you hear from the professor, but also where you meet your fellow study-group members, then go off to some other locale for serious, collaborative study. They noticed that students would try to congregate in the hallways. Why not push that trend further? they wondered. The solution they hit upon was to create "almost" rooms along the major circulation path. Doorless alcoves with whiteboards, plug-ins for laptops, and flat-panel monitors for collaborations. Students could reserve these "almost" rooms for blocks of time with Room Wizard, an electronic room reservation device jointly created by IDEO and Steelcase. The alcoves offered the openness of a boulevard café and the privacy of a library nook. The concept was an instant hit with students.

Look around your organization. Do you have a lot of dull hallways? Is there something simple you could do to increase spontaneous interaction and relaxed collaboration? Is there a corner or spot where a café table and chair might spawn a natural hub?

Tear Down Your Temples

When you create new work environments, you're bound to bump up against egos. People with power generally look upon offices or buildings as reflections of their achievement. A tour of Rome or Paris will remind you of that fact, as will a visit to Times Square or Washington, D.C.

So why should an executive, business leader, or other professional behave any differently today? Because the world has changed. Because what you can *achieve* matters more than what you can *accumulate*. Good space design leaves your business room to adapt and grow. It recognizes your strengths and reflects the underlying flexibility of the enterprise. Unchecked pride is rarely healthy for your organization. It sends the message that you care more about your status than your process or your customers.

We once worked with a group of cardiac surgeons at a prestigious hospital who had ample reason to be proud of their work. They saved lives every day, earning the respect of their peers and the community. They knew exactly what they wanted—and what they felt they deserved in their new space: A separate entrance exclusively for cardiac patients. A temple to their practice.

We found ourselves wondering, How many entrances does a hospital need? Then, we listened to the surgeons and the patients. We did observations at the main entrance, in the lobbies and in patient rooms. And we enlisted the surgeons' participation in a process of empathic design. What's it like to be in the patient's shoes? we asked. What are you feeling when you arrive for open-heart surgery? You're anxious, to say the least. Does it make you less anxious to know that you enter through a separate entrance, one reserved for those with health problems so threatening that they require surgery? What might that do to your psyche?

We let the surgeons be the final arbiters. They continued to like the idea of a separate entrance, but once they'd finished the empathy exercise, they saw other possibilities. Instead of the separate entrance, we suggested a grand escalator to the second floor.

Has your success spawned a proud temple? Or does your space genuinely invite partners, clients, and customers to visit?

The design offered something for heart patients (not having to walk) and for surgeons (the elevating distinction of a unique interior entrance). The escalator celebrated the professionalism of the department without glorifying it and without running the risk of unduly frightening arriving patients.

Odds are you're not a surgeon, but there's a good chance you're proud of your work. Look around your office and ask yourself if your space sends the right messages. Has your success spawned a proud temple? Or does your space genuinely invite partners, clients, and customers to visit? Does it give you room to work independently and in fruitful collaboration with others?

What might you do to make it a space that captures your imagination?

Paper Walls

Every firm has to strike the right balance in the trade-offs between private and collaborative space. At IDEO, the collaborative nature of our work demands that the majority of our space embrace teamwork. We favor an open plan with lots of project neighborhoods and shared spaces for team meetings.

Some cultures, of course, want or need more private space. For instance, we know of a prominent company that recruits high-level software engineers partly by offering the perk of a private office. The original idea was to blend privacy with collaboration by using glass walls in the offices, maintaining visual contact among team members while giving some acoustic isolation. Newly hired programmers loved the combination of privacy and light. Once they took occupancy, however, they quickly papered over the windows, turning the semiprivate offices into, well, paper caves. And so went any hopes for a space open to collaboration. Rumor has it that the company recognizes that their space is a barrier to good teamwork but can't work up the nerve to lead the programmers out of their papered caves.

I love this story because it demonstrates what can happen when Set Designers ignore powerful human impulses. The software specialists were assigned offices that lent themselves to privacy. Why should anyone be surprised that they improvised and pushed their offices even more toward the private realm?

Remember that *people* are the X-factor in space. Never forget how people might react to space—whether it's collaborative or private. They're the ultimate arbiters.

Follow-Through

Most Set Designers intend to create effective spaces, but sometimes there is a gap between intentions and reality. On a project with a West Coast university, for example, we found precious little shared public space that was useful to students. When asked about the apparent oversight, one university administrator pointed to an empty courtyard. The plaza-sized space was clearly meant to be a public resource. It probably looked good in the original architectural renderings and floor plans. But the current design had failed the students. There was no reason to visit or pass through the plaza. Yes, it *looked* like a public space, but the IDEO team never saw a single person make use of it. No casual conversations. No impromptu study group meetings. It was dead space.

I'd argue that practically every company has some dead space. You may have designed or forecast a worthwhile purpose. But as people adapted and moved on, the place was left behind like a ghost town. You can renew these forgotten places by seeding some group activity there, turning them into project rooms, or interactive display spaces, or spots to post information about relevant emerging trends. At one organization, we found that simply placing an espresso machine in an unused space turned it into a hot spot.

Here's the good news for Set Designers. You don't have to immediately come up with the be-all-and-end-all purpose of these overlooked places. Once people discover them, they'll likely find new uses for the spaces and they'll gradually take on a whole new life. Stewart Brand, creator of *The Whole Earth Catalog*, argues that buildings adapt or "learn" over time, as their inhabitants shape and mold them to their changing needs. One of the challenges of the Set Designer is to create spaces that will learn gracefully and quickly to support the kind of dynamic teams that spark continuous innovation.

CREATING AN INNOVATION LAB

Innovation needs a place to flourish and grow. A place where teams can meet, discuss findings, hash out prototypes, and present their work. A place people running innovation projects can call home. I have noticed that the creation of an innovation lab at corporations often coincides with a surge in innovation. It's not necessarily cause and effect, but there's still a noticeable correlation. We've seen companies build successful innovation labs located around their corporate campuses, not out in the proverbial skunk works. Getting your innovation space just right for your group will probably take a little trial and error. Start with something simple, and plan on learning along the way.

Innovation Lab Basics

Once you get started, your innovation lab will develop a life of its own, so the first steps are the hardest. Here are a few hints that might help if you're doing this for the first time:

○ Make room for fifteen to twenty people, even if the core project teams will be small. You'll want to share results (and even work in process) with lots of your colleagues, and that's best done in the innovation space. Don't make it too small for good group presentations or learning workshops.

○ Dedicate the space to innovation. Your creative efforts need to live on without scheduling or moving. Workplace experts call that "persistence of information," but I think of it as just keeping up the group momentum.

○ Leave ample wall space for sketch boards, maps, pictures, and other engaging visuals. Don't use delicate surfaces or precious materials that would inhibit maximum creative use of all vertical and horizontal surfaces. I recently visited a beautiful new corporate learning center that would be the envy of many innovative teams. But when setting up for a workshop, we were hampered in putting up large interactive posters by a "no tape on the painted surfaces" rule. The pristine beauty of the space had inadvertently become a barrier to idea flow. Try not to let that happen in your innovation space.

○ Locate your lab in a place convenient to most team members, near enough for even part-time team members to drop in on a moment's notice but far enough away so they can't hear their desk phone ringing.

○ Foster an abundance mentality. Stock the lab with an oversupply of innovation staples: prototyping kits, Post-its of every size and color, masking and duct tape, foam core and poster board, scissors, X-Acto knives, blank storyboard frames, fat-tipped felt markers for drawing up the broad outline of an idea, research kits, and so on.

The Power of Place

Perhaps you don't think of yourself as a Set Designer. Maybe it's just a daily struggle to keep your office functional. Well, don't worry. Even little adjustments can make a difference. My friend Tom Peters taught me a simple but revealing lesson about space. If you want to make something important, put it where you can't avoid it. Give useful things a place of prominence.

Like many of us, Tom owns one of those big, thick, unabridged dictionaries, which he kept for many years high on a shelf in his home office. Once every two or three months, he hauled the massive tome off its perch and laid it on his desk to look up a word. Then, one day at an antique store in Palo Alto, he found one of those classic maple library-style dictionary stands that hold a reference book open and ready on a moment's notice. Now, Tom reports, he looks up two or three words a *day* when he's working at home. The simple act of leaving his dictionary open and on a more convenient level instead of closed up on the shelf caused him to look up words nearly a hundred times more often.

What could you move that would make it more important, more approachable? Can small shifts in your space change behavior? Retailers understand this principle instinctively and know, for example, that an item on the front face of a supermarket aisle—the endcap—sells faster than if it's on a random nondescript shelf. As an author, I know all too well the value of being on that featured "current releases" table in front of the store. Though it may not be as easy to measure success in sales per square foot when you fine-tune your office environment, I believe the payoff can be even greater. Think about what you want to emphasize, like collaboration, creativity, or perhaps your passion for a particular field. Then experiment with some subtle changes. Maybe something as simple as cleaning out unused files or books—to make room for what you really need. Making room for an associate to pull up a chair. Creating a bulletin board to display wild ideas and articles that intrigue you. Putting the water cooler (or coffeemaker) near your desk if you want company and conversation. Give it a try and you may find that small changes can make big improvements.

> To Frank Boyden, the "important" place for himself was not the corner office on the top floor. It was right in the midst of the people he cared about most.

As for moving important things, what could be more important than *you*? It pays to move yourself around a bit, not just to keep things fresh but to seek out settings that better suit your workstyle. The upward career path of most successful executives leads them to spaces more and more isolated from their teams. Isolation may bring benefits of privacy and status, but sometimes we're better off right in the thick of things. Consider this remarkable example from the world of academia: Frank Boyden, headmaster of the prestigious Deerfield Academy in Massachusetts, had an extraordinary method for staying in touch with his students. Far from being some remote administrator, Boyden came up with an ingenious method of making sure he saw every one of his pupils every day. When they built a new school, Boyden had the architect widen the hallway in one spot so that his desk could fit into a little island amid the stream of students.

The constant interaction provided him with a well-informed intuition about individuals and the student body as a whole, giving him an uncanny ability to spot small morale problems before they got to be big ones. Boyden earned a reputation as one of the leading headmasters in America, and he was the subject of John McPhee's book *The Headmaster*. To Frank Boyden, the "important" place for himself was not the corner office on the top floor. It was right in the midst of the people he cared about most.

I know what Frank was thinking. Not long ago, I moved into a new space on the second floor of IDEO's headquarters building with several other senior members of the firm. It was the kind of setup most executives crave. Window view, ample space, a semblance of acoustic privacy. The catch? My office was so isolated, I seldom saw anybody. What did we do? The executive team gave up our window seats and compressed the size of our offices by nearly 50 percent. No view, no privacy, but I was suddenly just a jump shot from where everybody picks up their daily mail. Five paces from the spontaneous conversations that crop up in the central kitchen where more than a hundred people stop by to get their coffee, bagels, and cereal. Within earshot of my brother David's and CEO Tim Brown's offices. Just around the corner from our broad racks of eclectic magazines. I went from seeing nobody to seeing practically everybody in the office — including most of the visitors who drop in from our locations around the world.

Now, during the course of the day, I pick up snippets of informal conversations that give me a real sense of what's going on at IDEO. I feel more connected and in touch with my fellow team members. So the next time you close your office door or tuck in behind a cubicle, consider whether it's time to get out into the stream of your organization. How might you make the place you work more a part of your company?

Don't let space be the weakest link in your business. Tap into the power of the Set Designer and make your workspace and office one of your most versatile and powerful tools. Great space can enhance morale, improve recruiting, and even increase the quality of your work. Put people in great work environments and you may find they feel like staying a little longer or putting in a little extra effort. Maybe your group is poised for greatness and the right space would be just enough to help you cross over the line.

Set the stage for your team and you may be surprised how well they perform.

CHAPTER 9
The Caregiver

Think one customer at a time and take care of each one the best way you can.

—GARY COMER, FOUNDER OF LANDS' END

The Caregiver is the foundation of human-powered innovation. Dedicated and inspired doctors and nurses represent the purest form of Caregiver. Think of your best personal experience with a good physician, how they took care of you in a professional way and at the same time nurtured you in a style hard to match outside the bounds of family. Maybe you had the kind of mom who took extra-special care of you when you skinned your knee or caught a nasty cold. That's exactly the sort of Caregiver we all hope for when we go to the dentist, the local clinic, or our personal doctor. And it's a model that translates well into all kinds of businesses.

Medical care is a powerful metaphor for the ultimate in customer service. Not the administrative difficulties or insurance paperwork, but the moment when a gifted doctor or nurse is actually delivering care, the moment when you are one-on-one with a medical professional giving you their undivided attention. A skillful doctor draws on all their training and experience to make a diagnosis, deliver therapy, monitor your status, or simply make you more comfortable. In that moment, you feel like you are the center of the universe, as they make critical decisions about how to treat your illness and look after your well-being. The best Caregivers exude competence and confidence—evoking that classic phrase "great bedside manner." They have well-reasoned answers to your questions and help smooth away some of your worries. They may not need to spend a lot of time with you to work their magic, but their presence is very reassuring. And they leave you with the calming sense that they have used their Caregiver skills to set you on a better path.

We all crave a good Caregiver. Why else would personal trainers be so popular? Why are some hairdressers in such high demand? Think of that great waitress or restaurant owner who shines attention upon you— the Caregiver who makes you feel you are the only customer in the room. Caregivers have empathy. They work to extend the relationship. They show rather than teach. And they are very good at guiding your choices.

Caregivers take extra pains to understand each individual customer. Why? Because the best care is geared to personal interests and needs. Just as a good hospital combines efficiency with warmth, a good service accomplishes the task at hand while treating you as a person throughout the process. Think of what irritates you when you're waiting in line at the cash register. A salescleark leaves you in the lurch by going to someone else first, or juggles a mobile phone call midtransaction. Small variations can be the difference between making you feel like you have "special customer status" or, conversely, like a "customer unit" being processed by a bland service machine. Caregivers know that many services can be made simpler and a lot more human.

To understand this principle, let's take a trip to the hospital emergency room—one of the front lines of caregiving. Robert Porter, CEO of SSM DePaul Health Center in St. Louis, was one of the first health care executives to make the connection between IDEO's human-centered design work and the uniquely challenging world of health care services. Among other topics he asked us to consider was the patient's experience of the emergency room. (Nearly all of our health care clients now call it the emergency *department*, reflecting the fact that it's usually a complex enterprise spanning more than one room.) We did our usual anthropological observations—one IDEOer pretended to have a broken leg and somehow managed to videotape his entire hospital experience. We saw abundant evidence that people are under tremendous stress when they arrive at the emergency department. We also saw that the department can be extremely confusing and frustrating on first encounter. What did we do? We created a simple map for the staff to give to every incoming patient—the seven steps of your emergency visit, beginning with seeing the triage nurse, navigating insurance, being evaluated by an RN, being seen by a doctor, and so on.

One thing we noticed was that first-time hospital patients have trouble keeping the staff straight. There are nurses, administrators, technicians, doctors, radiologists. A few years earlier, our San Francisco office had made up "baseball cards" for its IDEO team to help new clients keep track of who's who. Building on that idea for the DePaul project, we suggested using the baseball-card concept for their Caregivers. Unmistakably different from traditional business cards, the vertical-format cards featured photos and personal backgrounds—who they are and what they do. The cards humanize the staff and build familiarity. The idea seemed like a simple way to personalize what can otherwise feel like a lonely experience. Stick people in a very stressful situation, and a little more information, guidance, and navigation make them a lot more comfortable.

No, IDEO and the DePaul team didn't change the actual clinical procedures at the point of care. We didn't change the doctoring. Yet DePaul has told us that the procedural and culture changes set in motion on that brief project have yielded a better patient experience. What's the

Patients feel better cared for when you guide them through the process.

difference? For one thing, we've spelled out the steps. We've explained the process. Where you begin. What forms need to be filled out before you see the doctor. At what point the doctor will decide to admit you to the hospital or send you home. It's the difference between the parts and the whole. I think patient-centered explorations like this one can help hospitals make the transition from treating Mrs. Henderson's appendix to treating Mrs. Henderson.

Of course, most companies are not in the business of practicing medicine. But the lessons we've learned about breaking down complex experiences into manageable steps—into simplifying and humanizing care—translate directly into many other industries.

Personal Taste

In the retail world, for example, consider the often frustrating experience of buying a bottle of wine that suits your tastes. For starters, the vast array of choices at even a medium-size wine shop can be overwhelming. The wide price range can be both scary and perplexing. And in some shops, asking a naive question seems like an admission of weakness.

Joshua Wesson, cofounder of the Best Cellars wine shops, seized upon a simple truth. While lots of folks enjoy drinking wine, the mysteries, confounding rituals, and downright snobbery surrounding the industry often put people off. I recently met Josh at an event for Australian winemakers and was impressed at the level of service and care he's created in building a better wine-buying experience. He's built his entire business around making wine buying more personal, simpler, and more fun. He likes to say, "We know wine. You know what you like." And his whole store experience is about combining the two for a happy result.

His wines are budget-priced from $5 to $15, and he's gotten excellent reviews in publications from the *New York Times* to respected wine magazines. He has neatly dismissed the need for customers to be experts in vineyards, vintages, or varietals. Best Cellars has simplified all those factors into four taste categories for white wine—fresh, soft, luscious, and fizzy—and four for red—juicy, smooth, big, and sweet. Adjectives also help describe the categories, such as "concentrated, powerful, and satisfying" for the "big red" wine category. Carefully crafted graphic design and interior architecture also help guide the prospective buyer. Color-coded icons let you know where to find a luscious white or a smooth red, and the wines are staged and backlit thematically like an upscale boutique (thanks to set designer David Rockwell, who helped create Best Cellars' flagship store in Manhattan).

Visit Best Cellars' Web site and you can get more interactive help. For example, you can take a taste quiz, answering such offbeat questions as "What's your favorite syrup with pancakes?" to determine whether you're likely to prefer a "fresh" or "soft" white. Stores offer three weekly tastings, encouraging customers to experiment. Inclusiveness is the goal. Best Cellars' central promise is "We've tried to remove any obstacles between you and your enjoyment of wine," and I wonder if there isn't a lesson there for businesses everywhere. Identify every barrier that keeps people away from your offerings, especially for first-time customers. Then systematically tackle each one, using a combination of simplicity, clear communication, and customer-centered design.

When I met Josh, he talked about some of the wine pairings he's done at stores, upholding the industry tradition of matching foods with complementary wine flavors. But Best Cellars' pairings go way beyond the ordinary. While traditional pairings almost always seem to match wines with fine cuisine, Josh told me he's happy to pair his wines with any food you're likely to eat, including a peanut butter sandwich or a Big Mac. Josh firmly believes it's good business to deflate the snobbery that riddles the wine industry.

The entrepreneur hosted a wine tasting that evening, which for me at least proved the main point of his stores—that for most everyday occasions, a well-chosen $15 wine can be pretty darn good. After tasting two smooth reds, he asked us to vote which was our favorite. Roughly an equal number of us voted for wine #1 versus wine #2. After revealing that #2 was in fact the more expensive wine, he asked those of us who selected it whether we liked our chosen wine 20 percent better. Yes? How about twice as much? Three times as much? "I hope you liked it six times as much," Joshua then told us, "because it

BEST CELLARS
S E L E C T I O N

fizzy	sparkling wines
fresh	light-bodied white wines
soft	medium-bodied white wines
luscious	full-bodied white wines
juicy	light-bodied red wines
smooth	medium-bodied red wines
big	full-bodied red wines
sweet	dessert wines

Best Cellars takes care of its customers by rejecting snobbery and helping people find wines they like.

cost six times as much as my wine—$90." I tasted the first wine again—poured from the $15 bottle—and realized it really wasn't so different from the far pricier red.

So what has Best Cellars achieved? They're making people feel more comfortable about stepping into a wine store. They're making customers feel more confident about their choices. Best Cellars is like a club that always welcomes new members. That's what the Caregiver role is all about. Josh Wesson has figured out how much care and service goes into bringing you a very good experience.

Josh has fine-tuned his successful offering for affordable wines, but the same principles work at the high end of a market. A few years ago, one of my retail-savvy friends pointed out that Tiffany ads now often mention a price for their elegant jewelry—which is unusual in a luxury category—as in "necklaces, from $100" or "Tiffany engagement rings, from under $1,000." "Know why that is?" my friend asked. I didn't. It's to lower the fear factor. It's to let you know that the clerk at Tiffany will not laugh at you—hopefully won't even look down her nose at you—if you tell her that's how much you have to spend. Yes, of course, you might leave the store paying a bit more than that. Maybe a *lot* more. But hats off to Tiffany for getting the word out that you don't have to arrive in a limo to feel comfortable shopping there. And if you are still scaring away your entry-level customers with an air of exclusivity or needless complexity, don't be surprised if they become someone else's customer and never come back.

Slow Hands, Fast Car

Sometimes we can learn about Caregiver service from the strangest places. One great source of inspiration is what we call "experiences on the edge."

IDEO has always attracted unique personalities with unusual backgrounds, affording us a window into some "edgy" worlds. Take Juan Bruce, for example. On the surface, he's a fairly typical IDEO employee with a product-design degree from Stanford. But at age thirteen, the young car enthusiast penned a one-page proposal to his parents for the high-performance car he wanted to buy—winning over his skeptical father and mother at a time when he was still three years away from being "street legal." The car he chose was an aging BMW, and Juan was soon honing his mechanical skills by taking it apart and putting it back together again. By fifteen, he was turning ovals on a racetrack (which is legal, he tells me), and by seventeen, he was competing in races.

While in his early twenties, Juan became an instructor for the famed Beverly Hills Motor Sports Club. Juan's clients have ranged from a cross section of CEOs to celebrities like TV star Matt LeBlanc and heavy-metal rocker Axl Rose. The race cars these guys truck to the track are top-of-the-line Ferraris, BMWs, and Porsches. They all want to go fast. Really fast. Juan risks his life in the front seat, at speeds of more than 150 miles an hour, with perfect strangers.

Juan is still on the sunny side of thirty, considerably younger and definitely less famous than many of his clients. So what are the secrets of giving advice to the rich and famous, people who are accustomed to being in charge? For starters, they want a driving coach, not a fan. Juan treats them like people—admittedly, exceptional people. Diplomacy is key, says Juan. He does his best to take the wheel the first couple of laps so that he can effortlessly show, rather than talk about, good driving. But if they're not interested, he takes his chances riding shotgun in the passenger seat. That's where he demonstrates his patience. Rule number one is they don't want a lesson. Certainly not a teacher. Winston Churchill used to say, "I'm always ready to learn, though I don't always like being taught," and many of Juan's clients feel the same way. Maybe we all do. So Juan aspires to the antithesis of the traditional teacher-student model.

He tries to deliver learning without visibly teaching. For clients looking for the ultimate in high-speed excitement, Juan represents the quintessential Caregiver. And the service he delivers has no room for error.

Since virtually every CEO and celebrity client loves putting the pedal to the metal, Juan must use subtlety and tact to impart a key Zen-like principle of race car driving: "Slow hands, fast car." Fast drivers are slow in the car, says Juan. The celebrities want the high of feeling the speed, but the truth is there's almost a Zen calm to achieving it. Smooth gear shifting and slow turning translate into increased velocity, while abrupt or jerky movements can be downright dangerous at 150 miles per hour.

> Winston Churchill used to say, "I'm always ready to learn, though I don't always like being taught."

Juan gently encourages his clients to look beyond the turn, to think ahead. To anticipate. Not to come in too early or fast into a curve. By the time they enter the turn, Juan knows whether they're in trouble. "We went in too fast," he'll say as their high-performance Pirellis barely hold the turn at over 100 miles an hour. "Be nice and easy coming out so we can make it." That's the part that amazes me. Juan resists the impulse to grab the wheel. My high school driver's ed instructor often lost his nerve at 35 miles per hour, but Juan Bruce remains tranquil at race car speeds while his life is in the hands of celebrities not celebrated for good judgment. Even Juan admits to having had days where he had second thoughts. Like the ride with Axl Rose. "I was scared," says Juan. "But he turned out to be very calm."

I can't speak for Axl Rose, but I suspect it's a case where the clients come to reflect the coach. Great driving instructors don't have many bad students. Whatever the abilities of their clients, they manage to bring out their best. They customize their approach and style to each individual. They offer people an experience so exciting and seamless that clients momentarily forget that they're being served.

You probably aren't in the race-car industry, but there's a valuable lesson here. The higher the level of service, the more expert and transparent the customer service becomes. Caregivers shift from telling to showing, from serving customers to helping individuals. The Caregiver is less about lording knowledge over customers and more about sharing insights. They're more like a mentor—with a small *m*.

Extreme Service

Juan Bruce supplies his clients with a service offering at the highest level. But what about the other side of the equation? What can we learn from people with remarkable *demands*? What insights might they offer to Caregivers working with ordinary mortals?

A few years ago, for instance, we got a call out of the blue from Lufthansa Teknik, a specialty division of Lufthansa known for its incredible custom work for private and commercial jets all over the world. Our assignment? Fly to Germany to help them make a touch-screen remote control for passengers on high-end private jets. Since there's basically no such thing as a "generic" private jet, we focused on pleasing the launch customer: one discriminating multimillionaire with his very own custom Boeing 737. It's the same basic airframe made popular by low-cost carriers like Southwest and JetBlue, but you won't find any little bags of peanuts on board this jet—nor even seatback videos, since the seats are *way* too far apart to use that familiar commercial setup.

The client wanted something as simple as an iPod yet able to control practically every device in the most luxurious jet cabin imaginable. The sheer number of desired functions made the interaction design hard to simplify. Early on, we put more than thirty machines up on a chart—and the client wanted to control *all* of them with a single wireless device. I have trouble enough using one remote to control both the TV and the DVD player. In this case, the client wanted a single remote to do everything from adjusting the temperature in the cabin to moving externally mounted cameras.

We quickly realized that it was about more than creating a great remote. We had to orchestrate great experiences at the touch of a button. Experiences that could be selected like a performance, such as launching the first song in a favorite MP3 playlist, or dimming the lights and lowering the shades to view the landscape from the outside cameras. Conversations with the celebrity client, cabin staff, and flight

Lufthansa Technik pampers its high-end customers with a combination of the latest technology and ultimate service.

engineers helped steer us toward a central premise that I believe makes sense for many kinds of more ordinary control boxes. The airplane remote operated on two levels. Basic or synchronized experiences were controlled with a touch of a button. Then there was a deeper level, where custom settings (and combinations of settings) could be created. Meanwhile, the device also managed to control a widely diverse collection of machines, equipment, and digital devices over one Ethernet network. My favorite electromechanical feature is a magnetic docking system that holds the remote securely in its mount during takeoff, turbulence, and landing. It's wired into the jet's seat-belt signs, so it docks automatically every time the pilot advises passengers to stay in their seats.

Much of the credit goes to that first elite client, who not only drove IDEO and Lufthansa to create a remarkable device but also spent six patient hours one day testing virtually every element of the remote in his parked jet on a runway. The client wanted his remote to be instantly recognizable as something patently different from the remote you'd get with your home TV. Porcelain was suggested as one wild-card material, but was rejected as too delicate. Instead, machined Corian gave the remote an elegant but sturdy shell, while plated aluminum etched with graphics lent a handcrafted aesthetic. The luminescent "feet" on the underside of the remote control glow when it's placed on any flat surface—making the remote easy to locate within a dim cabin. Clients with fractional ownership (using the jet on a time-share basis with others) can take their favorite settings with them with a memory stick, so the system remembers their preferences each time they return.

Lufthansa's "nice" system—Networked Integrated Cabin Equipment—won Germany's prestigious Red Dot Award for excellence in design, and one of the devices now resides in New York's Museum of Modern Art. It's also part of Lufthansa's luxurious in-flight entertainment offerings. It's a great example of how empathic Caregivers can transform products or services into experiences.

A Great Fit

Caregivers understand that service innovations come in all shapes and sizes. But when you're stuck in a rut, practically nothing seems to work. At some companies, the first line of defense is that "our industry is different." They're thinking, "Sure, that idea may have worked for FedEx, Starbucks, or GE, but it wouldn't work in *our* company." They fall into the familiar trap of thinking there's no room left for change in their business. "Our industry is very traditional," they may say. "People would never go for that."

One answer to the Caregiver "idea drought" is a little shoe store named Archrival in Mill Valley, California, that breaks out of the traditional mold. Located in an open strip mall in Marin County, Archrival doesn't look very different from other sports-shoe outlets, so there's little hint that something remarkable is taking place inside. But hundreds of athletes and coaches swear by Archrival. Cofounder Peter Van Camerik learned long ago that while his featured product may be athletic shoes, what he really sells is caregiving—a seamless blend of service and expertise. Archrival prides itself on selling some of the best athletic shoes available and knowing the subtle differences that make one pair better suited for a certain foot type over another. The staff explains the nuances of high-tech shoe design better than Apple store employees talk Macintosh.

Peter asks a customer to stand in their stocking feet so he can quickly gauge whether they have a neutral or high arch or just plain old flat feet. He then observes their walk to see if they've got a hitch in their gate or other quirk. He measures a customer's feet and asks how much they run, play tennis, soccer, basketball, or walk. Then he generally brings out a single pair of shoes (his informed selection is key to the service) and watches you walk or jog around the store. Usually they're a comfortable fit, and you gladly plunk down your Visa card. Peter trains his staff not to sell a shoe until they've got "the story of the feet," the biomechanics. "Once you understand the story," says Peter, "you can prescribe the right shoe."

Mothers and fathers bring in their children for playground sneakers, soccer cleats, and everything else (several parents actually send their kids in alone with a credit card and a note for Peter). Some customers report that they haven't bought shoes anywhere else for more than a decade. "I walk in and tell Peter I need a new pair," says a loyal customer. "He brings out the shoes and they fit beautifully. I don't ask the price." Contrast that to the nearest competitor, a discount sports store. You're lucky if you can find a clerk to bring you a couple of sizes. When asked the difference between two shoes (one was for walking, the other trail running), a dumbfounded clerk told a would-be customer, "This one has more tread." Peter says many of his sales are to people on the rebound who have had bad experiences with discount shoes.

What's Archrival's edge? Peter is a natural Caregiver who recognizes that great service is about knowledge and empathy. He doesn't try to push the most prestigious brands or the most expensive models. A former competitive tennis player who suffered foot and back injuries, Peter sells a great fit in the right shoe. More than two decades ago, Peter called on the major podiatrists, orthopedists, and physical therapists of Marin and San Francisco and impressed them with his expertise. Today he gets upward of twenty referrals a day from doctors and clinics—customers walking in off the street with a prescription for shoes. Peter even hired a jovial retired podiatrist to work in the store, helping customers get a good fit and enlightening them about the care and health of their feet.

So, what can you learn from a shoe store about the role of the Caregiver? Archrival spends practically nothing on advertising yet has an expanding business. Margins are excellent, since Archrival's expertise and extra service make discounting largely unnecessary. Most important, Archrival's accent on customer service begets repeat buyers who buy multiple pairs of shoes over the year and invite friends to check out the store.

If a commodity business like shoe retailing can become the nexus of a thriving, creative enterprise, virtually any industry can be improved—no matter how long it's been done the same old way. Whether you sell products or services or even serve internal customers within a large organization, the lessons of this small but inspired Caregiver have broad value. Maybe it's time you started walking in some new shoes.

Offer Customers a Safety Net

I've talked about how Caregivers get closer to their customers, how they guide their experience and build in comfort. But sometimes it makes sense to encourage your customers to explore a little as well, especially if you have a distinctive range of offerings. One of my kids' favorite nearby restaurants is California Pizza Kitchen, the company who successfully proved that customers have an appetite for pizza toppings way beyond mushroom, sausage, and pepperoni. CPK has an eclectic menu that includes pizza toppings like Thai chicken, shrimp scampi, and Peking duck.

At first, the "California" moniker fooled me into thinking that it was a regional chain, but I have since seen their restaurants everywhere from Singapore to Vegas, and their revenues are pushing $400 million. In a fun twist on the time-honored money-back guarantee, California Pizza Kitchen has what they call their "CPK Menu Adventure Guarantee." "Be adventurous—try something new!" urges a side-bar on every menu. "If it doesn't thrill you, we'll replace it with your usual favorite." It's a smart piece of marketing and brand building when you think about it. First, if you are California Pizza Kitchen, the last thing you want is for your customers to be stuck in the pepperoni pizza rut. Sure, the chain restaurants make a good crust and use all fresh ingredients, but if all you ever order is pepperoni, you are a very vulnerable customer. As they say in the marketing biz, you have low barriers to exit, because practically every pizza shop on the planet has a pepperoni pizza. But if they can get you to fall in love with Kung Pao Spaghetti or Tricolor Salad Pizza, then you're much more likely to become a loyal convert, because who else are you going to call for those exotic offerings? The San Francisco yellow pages lists more than a hundred pizza places, but offhand, I know of only two that serve Thai chicken pizza—the two local outposts of CPK.

Let's not forget the safety net. Not only does the "menu adventure" allow you to take a walk on the wild side, but you also get the safety of knowing you can return to your traditional "comfort food" if it

California Pizza Kitchen dares its customers to explore exotic menu alternatives, while still giving them a safety net.

doesn't work out. The idea of experimenting safely is a powerful lure and has applications for businesses ranging from theme parks to variable annuities. Give your customers a chance to experience a broader range of your offerings without abandoning their safety net and they will find a way to reward you. Since food costs in a well-run restaurant typically run only 30 percent of total costs, CPK's guarantee not only builds customer loyalty but also costs the restaurant only a third as much as the traditional money-back guarantee. It's a service innovation that is rewarding for both parties.

How might you offer your customers or clientele a chance to experiment with your offerings in a risk-free environment?

THE CAREGIVER'S GUIDE TO GREAT SERVICE

1   Curate the Collection. In most product or service categories, customers are overwhelmed with too many alternatives and too little clarity about the right choice. Which college would be best for my children? Which mobile phone (and service plan) would best suit my lifestyle? Questions big and small with hundreds of answers to choose from, but too little trustworthy guidance on which is best for me. Customers need your expertise and knowledge to sort through all the possibilities. Prune your offerings so you are offering the best of the best for your clientele. Provide a small selection of excellent choices and have a point of view about why you chose those few among the many. Starbucks takes good care of me with their "Artist's Choice" music played—and sold—in many of their 10,000 stores. There is so much music out today that it's hard for me to discover new material I'd like and decide what to listen to. But Starbucks offers to share a playlist that represents, for example, Norah Jones's favorite music, and that makes it an easy choice.

2   Build Extra Expertise. If your company becomes a trusted source of information or advice, you'll build a base of loyal customers. Archrival began by educating local doctors, clinics, and podiatrists on the importance of solid, supportive shoes. They even hired a retired podiatrist

to give customers priceless knowledge about their feet and shoes. If price weren't an obstacle, whom might you hire to give your customers fantastic service? Alternatively, what trusted information sources could you connect your customers to that would help them become more informed buyers?

3   Small Can Be Beautiful. Think café, bistro, or barbershop when creating a service or sales outlet (if people just want to chat, you're doing something right). Maintaining customer "intimacy" translates into a greater concentration of people, more anticipation and action. Sometimes a network of small sites can provide more surface contact with the customer than a single big location.

4   Build Relationships with Sustainability. Invite customers to recycle your products and you'll create a virtuous cycle of giving and receiving. Your customers will earn a sense of satisfaction from knowing that they are helping others. In the process, you'll have given them yet another reason to bond with your brand.

5   Invite Customers to "Join the Club." Loyalty programs have become an extremely powerful tool for Caregivers in industries like airlines and hotel chains, but the model is still extensible into many other industries and settings. You don't need to hire a management consultant to tell you that it's good business to identify your most loyal (or most profitable) customers and give them "special customer status," whatever that might mean in your business. Frederick F. Reichfeld estimates in *The Loyalty Effect* that a 5 percent improvement in customer-retention rates can yield a 25 to 100 percent increase in profits—across a broad range of industries. Oddly, rental-car companies seem to understand this principle but some car manufacturers and dealers do not, because they are still thinking about the individual sales transaction when they should be focusing on the relationship. For example, my father has bought *fifteen* GM cars in the past fifty years—and had them serviced by GM dealers—but each time he walked into the showroom, they treated him like a stranger. How many customers do companies lose with such benign neglect? Sign your best customers up, take good care of them, and let them be your brand ambassadors.

Prototyping Customer Care

The Caregiver mentality helps deliver better products and services in everything from a luxury jet to a restaurant or store. But what about in an international organization with thousands of locations and millions of customers? Developing innovative services for a truly mass market can be an overwhelming task. The human element looms large, as does the question of figuring out what works and what doesn't across a wide variety of situations and locations. Most companies in competitive markets must continuously create and provide services to customers—and even their own employees—to stay competitive. Yet few large companies boast a creative, rigorous process for coming up with new ideas, testing them, and implementing them throughout a large organization.

At financial services giant Bank of America, however, they've developed a unique and apparently successful methodology to prototype new service offerings. Not satisfied with just having one of the largest banks in the United States, CEO Kenneth Lewis wanted to also be recognized as the best. The bank formed a new group to lead the charge— the Innovation & Development (I & D) team—creating a framework for real-life prototyping with the bank's own branches as learning labs. BofA selected twenty branches in Atlanta as their test bed, based on the installed base of technology already there, the composition of their customer base, and the proximity to their new corporate headquarters in North Carolina.

Harvard Business School professor Stefan Thomke, who studied the bank's innovative program, reports that there have been hundreds of ideas run through the process and dozens of live experiments conducted. For example, in the initial prototype branch, hosts started greeting customers at the door, directing them to associates behind kiosks offering a variety of banking and financial services. The atmosphere became more like a boutique private bank than the branch of an industry behemoth. Customers could read about investment products and peruse financial magazines while relaxing on comfortable couches. They could check their accounts or access the Internet from computers at the Investment Bar. Electronic stock tickers and flat-panel television sets tuned to CNN provided entertainment and information to those waiting in line.

Each new exploration was carefully designed and executed to test out one hypothesis about customer care. In the world of academia, economists like to use the expression *"ceteris paribus,"* a Latin term for the assumption that all other variables remain constant. But in the practical, commercial world where BofA started testing new customer-service concepts, all other variables *never* remained constant (that's why they call them variables). So the Innovation & Development group had to control for constantly fluctuating factors like weather, staff turnover, and seasonality by conducting tests at multiple locations at different times. Some of the most interesting tests had to do with managing perceived wait times, which are sometimes easier to influence than actual wait times—and possibly more important, because in the Caregiver's world, the customer perception of what happened is paramount. As Thomke reported in the *Harvard Business Review*, BofA discovered a hinge point at about three minutes of waiting in line, after which perceived waiting time raced ahead of actual time, causing customers to overestimate their wait. Experiments showed that it was possible to minimize this effect with video monitors that entertained and distracted customers as they waited. And since the bank knew from previous research that every one-point improvement in their customer-satisfaction index translates into an extra $1.40 in revenue per household, they were able to calculate a return on innovation for the new wait-perception concept.

Bank of America has made a significant investment in prototyping services aimed at better customer care. It may not be a perfect program, but the conscious effort—the creation of an innovation team, the selection of prototype branches, and the systematic testing of service improvement concepts—sets BofA apart. And as I write this, there's evidence that the bank's renewed focus on the Caregiver is on the right track. The *New York Times* has noted that unlike most banks, BofA now treats its bank branches as "stores" and has succeeded in training and promoting managers "who interact openly with customers on the bank floor." They've wisely put a lot of focus on identifying and pleasing "delighted customers," who tend to be loyal, interested in new services, and key to generating multiple referrals. And a *Financial Times* publication has rated the bank as "Bank of the Year in the U.S." two years in a row.

Could that service-testing concept work in your business? There's certainly nothing that limits it to retail banking. For example, we worked with the prestigious Mayo Clinic to create their SPARC Innovation program, using a dedicated space inside the clinic to develop and test innovations for outpatient health care delivery. It has some parallels to the BofA prototype branches, but of course is completely different in content and in the nature of testing and measurement. The VHA Health Foundation says the Mayo SPARC program is "the first systematic 'live clinical laboratory' in the health care industry" focused on outpatient care. And we believe it will be a model for many future programs not only in health care but in other service industries looking to improve their customer care.

The Doorbell Effect

The BofA research touched on the subject of waiting—an unavoidable element in most customer journeys—and I believe that the way you manage those critical wait times can make all the difference in how your company is perceived. Think, for example, of the worst service you've had in the past six months. Did it, by any chance, involve lots of excessive or seemingly unnecessary waiting? Probably so. But since a certain amount of waiting is inevitable, you should at least find a way to make it more palatable.

No good Caregiver would leave their customer completely in the dark during that waiting period, but that's exactly what most organizations do. I call it the "Doorbell Effect," because it's like that uncomfortable lag between when you push the doorbell button and the door opens. Or *doesn't* open. You have very few clues during that awkward waiting period. You're behind a door, after all: opaque and almost soundproof. It's hard to tell whether your summons is bringing the occupant to the door or whether the doorbell has just echoed through an empty house. Worse yet, you may not even be sure that the doorbell rang. Chances are we've all experienced that awkward no-man's-land. When no one answers after what seems like a reasonable amount of time, there is an almost overwhelming urge to ring the bell again.

Do you wait a bit more? Ring one more time? Risk annoying the occupant? Knock loudly, just in case the bell is broken? Give up and retreat gracefully? The problem is not simply the waiting. It's the waiting on pins and needles, not knowing what's next.

The same thing happens with many service businesses. They leave the customer hanging, uncomfortable, and uninformed. The "hang time" may seem perfectly reasonable to the marketing team, but a couple of minutes can sometimes be a very long time when you're standing all alone on the porch. And even a little information at the right moment goes a long way. Think how much easier it is to wait for the elevator when you can see what floor it's on, so you at least know that it's working today and it's coming soon. Notice how much better it feels when the customer-service line tells you "your call should be answered within the next three minutes"? Even the sign at the theme park that warns you "average wait time from here forty-five minutes" is way better than having no idea.

People who still mail in their tax returns to the IRS experience the ultimate "Doorbell Effect" every April 15. If you're like me, you always make it just under the wire, which gives you a small feeling of satisfaction. But then you wait. Say you're expecting a $500 refund check and you're hoping to get it as soon as possible. A week passes. No sweat. Two weeks. It's the government, after all. Three. Four. When you get to six or seven weeks, you start to wonder where that check is. Is it lost in the mail? Worse yet, did the tax return even reach the government in the first place? That's the most unnerving part of the worry sequence set up by the Doorbell Effect. I've always found that limbo state to be very unsettling, and was extremely pleased to see the IRS usher in electronic filing to minimize the Doorbell Effect.

Some companies, on the other hand, have been pretty astute about the Doorbell Effect all along. They let customers know where they

I call it the "Doorbell Effect," because it's like that uncomfortable lag between when you push the doorbell button and when the door opens. Or *doesn't* open.

Beware the Doorbell Effect.

stand, reassuring them along the way. Netflix, the California-based movie-rental company that ships out 3 million DVDs to their customers every month, was sharp enough to build reassurance into their business model from the very beginning. I keep five Netflix movies out on rental at all times, some at my home and some in transit. A couple of times a week, I mail them the DVD I've just finished watching, then wait for its replacement (the next movie in my rental queue) to arrive. We're stitching three lag times together: the time it takes the Postal Service to deliver my envelope to the company, the turnaround time at the Netflix processing/distribution center, and the time it takes to ship the new DVD back. If each of those steps takes, say, one to four days, I'll be expecting a return shipment in three to twelve days, which leaves a lot of room for uncertainty. Netflix beautifully sidesteps the Doorbell Effect by chopping my uncertainty time into its three more manageable component pieces. I drop my DVD in the post box, and, within a couple of days, I get a quick e-mail from Netflix saying "We've received *Almost Famous*" (or whatever my latest movie was). Then, usually within twenty-four hours, they send another update saying "We've shipped *Lord of the Rings*. You should have it by Tuesday." I always know where I stand. I'm never left wondering for very long. And, best of all, if the movie ever does get lost in the mail, they simply replace it for free.

Unfortunately, Netflix remains the exception to the rule. Take a look around and you'll see customers still waiting expectantly out there. When I double-click on the icon for my Web browser, for instance, sometimes nothing happens for five or ten seconds. No hourglass, no progress bars, no "Please stand by." Just my computer screen staring blankly back at me. Left out in the cold, I double-click again, just to be sure, and in doing so, I make my situation worse.

Whatever you do, whether you sell products, deliver services, or provide information, odds are that your customers must spend some time waiting—whether it's for something to pop up on their desktop or show up on their doorstep. Caregivers don't leave customers out in the cold. So beware the Doorbell Effect. Give them a better idea of where they stand and they'll be more loyal to your brand.

Mediate, Don't Automate

One of the central premises of this chapter is that some of the highest levels of caregiving are inspired by a human touch. But when automated systems have become ubiquitous, how do you maintain that personal connection? What guiding principles might Caregivers keep in mind when it comes to Web-based or other computer-driven services?

Interestingly enough, low-tech services may help inform us about what's critical in a technically charged world. For instance, consider a relatively impersonal style of service most business travelers have experienced—ordering morning coffee at a hotel using the card you hang on your doorknob the night before. You take your pen and check off the options. Check here for decaf . . . for cream . . . skim milk . . . sugar, and so on. Most of the time you get your coffee, but I'd dare say it wasn't much of an experience. Compare that to Starbucks, where a far wider and richer range of options is laid out before you. Heightening the experience is the invigorating scent of espresso brewing and the sound of beans grinding. The barista is there to help you create your perfect coffee and, of course, offer you a blueberry muffin if you desire. The point is that, unlike checking off the doorknob card, you get the chance to play a more integral role in the process. It's not purely for operational reasons that they ask for your name at Starbucks. Doing so helps personalize the experience.

There's a value in maintaining human control and interaction—no matter the level of automation. Make sure that the people who use these new services have real, tangible interactions that enable them to fashion an experience or product they can call their own.

Just because full automation is possible doesn't mean it's necessarily desirable. Mat Hunter, a talented interaction designer who heads our London office, uses a simple expression that says a lot: "Mediate, don't automate." "Figuring out what you don't want the service or machine to do may be the most critical step," says Hunter. In other words, leave what's best done by humans to, well, humans. Too often there's an assumption that the software has to take control and resolve every problem. Better design, says Hunter, is to realize that you can't automate your way around every crisis—and that, in a high-tech world, sometimes the best services are those that use the wonders of technology to deliver even better traditional person-to-person customer service.

This is why no matter how high-tech the best hotels in the world become, there will still be someone at the front desk for you to talk to if you so wish. This is why the real key to improved customer service will be using sophisticated technical information to help smart and well-trained people deliver even better customer service. And why companies that struggle to solve too many problems for their customers—without inviting them to have a hand in the process—may be trying to do too much and offering too little.

The Cost of a Smile

Finally, there is one fundamental element of the Caregiver I don't expect to change anytime soon. Here's a brief, impassioned argument for an often-overlooked tool in the world of customer service: the smile. With personal and professional ties to Japan, I have traveled to Tokyo more than twenty-five times. On business trips, I've always flown United, but for family travel we started flying Japan Airlines in the early nineties. Boarding a JAL transpacific flight for the first time, I

Tap into the power of a smile. Your customers will notice the difference.

immediately noticed something different: Almost every single flight attendant smiled at us as we came onto the aircraft. What a concept, I thought. I wonder if United knows about this? Everyone thinks innovation is expensive, but how much can a smile cost? And, to my untrained eye, those smiles seemed both genuine and friendly, giving their company a slight edge in a very competitive business.

One organization that has figured this out is the Ritz-Carlton Hotel Company. Though the old-world elegance of a Ritz-Carlton hotel is not always my first choice, I am consistently impressed with the extraordinary service level at their properties when I stay there. I got a behind-the-scenes look at how service with a smile gets operationally reinforced early one Sunday morning at the Ritz-Carlton Philadelphia. I was scheduled to give a talk at 7:30 A.M. to an audience of association executives, many of whom—like me—had just flown in from the West Coast and found the early start challenging. The annual conference didn't begin until 9 A.M. that day, but this enthusiastic group wanted to get a running head start, so the Ritz-Carlton was hosting them for breakfast.

I arrived on the scene an hour before breakfast, and so I witnessed the pre-event pep talk from an energetic crew chief. In a room dripping with gilded chandeliers, amid tables adorned with their trade-mark blue goblets, three dozen Ritz employees amassed to prepare for the guests who would soon arrive. The crew boss started with an overview of the whole meal, then a detailed description of individual menu items and service elements. Just before he finished, he said, "This is an important group, and we want them to have a very good experience with us today. So do your best, pay close attention to their needs, and don't forget to *smile*. It puts them in a better mood, it puts you in a better mood, and it is part of our signature service."

Watching him motivate the team that morning, I felt a little bit of it rub off on me—and I wasn't even his target audience. In the process, I realized that though the tradition of "ladies and gentlemen serving ladies and gentlemen" is deeply embedded in the Ritz-Carlton culture, the company must continuously and powerfully reinforce those values. So though the smile is spontaneous and free, it is also the result of careful nurturing. It is part and parcel of the hiring, training, and culture of a company that sincerely believes in the value of friendly, professional service that epitomizes a Caregiver.

It may seem to be a small thing, but no serious Caregiver should overlook it. I daresay most of us (and most organizations) could do with a few more smiles.

CHAPTER 10
The Storyteller

The universe is made of stories, not of atoms.

—MURIEL RUKEYSER

The power of a good story has a few thousand years of history behind it. Storytellers have captured the rapt attention of their fellow humans for as long as there have been evening fires to tell tales around. Countless poets and bards sang the epic stories of *The Iliad* and *The Odyssey* long before Homer put them down in print 2,800 years ago. Shakespeare used his storytelling craft in the sixteenth century to spin history into literature, and he remains a global bestseller today (though he never got a piece of the movie rights). Even in the twenty-first century, popular filmmakers like George Lucas are wise enough to realize that a good myth is timeless, and his epic films, rooted in mythology, have grossed something approaching $10 billion, providing powerful evidence of the Storyteller's enduring value.

Brand-savvy modern business organizations also know how to tell a good story. They capture our imagination with compelling narratives of initiative, hard work, and, yes, innovation. They celebrate success and honor stirring recoveries. Whether we consciously realize it or not, businesses are constantly telling stories to their customers, their partners, and themselves. There's the story of a great collaboration, the story of a novel product or a full-bodied service—even the classic tale of a great venture launched in a garage.

Stories persuade in a way that facts, reports, and market trends seldom do, because stories make an emotional connection. The Storyteller brings a team together. Their work becomes part of the lore of the organization over many years. Storytellers weave myths, distilling events to heighten reality and draw out lessons. Going beyond their oral tradition, modern Storytellers now work in whatever medium best fits their skills and their message: video, narrative, animation, even comic strips. They help inspire other Storytellers to spread the word. Most important, Storytellers make heroes out of real people.

David Haygood, our head of business development, is a natural Storyteller. At first, I thought it was just that Haygood had a more exciting life than the rest of us. It's true that he does volunteer work inside prisons with convicted felons and goes wilderness camping with Navy SEALs. I've heard him recount his swords-to-plowshares story of how, as a draftee in Vietnam, he and his buddies would turn a cold C-ration into a warm pizza by mixing in some extra ingredients and heating it with a tiny pancake of C4 explosives. And he's shared lots of stories about dramatic successes—as well as failures—in a business context. But then one day, while he was telling a really entertaining story about a Monday-morning status meeting at Specialized Bicycles, it dawned on me that it wasn't just his material. Anyone who can turn a status meeting into a riveting tale is a master at the art of storytelling. And it's a very endearing trait.

Timeless Stories with a Purpose

What's another essential truth of storytelling? The creating and telling of myths is part of human nature. It is bigger than any individual organization. Even nations have their time-honored myths—stories strongly associated with a person or an institution that reinforce a cultural value. On a family trip to Boston last year, I was reminded of the story of Paul Revere's midnight ride, triggered by a signal etched into our collective memories: "One if by land, two if by sea." The tale is not only a history lesson studied by most American schoolchildren, but also a story that reminds the listener that one person can make a difference.

There are similar examples in every country and culture in the world, and they are not all about heroes in the traditional sense. Japan has the story of a faithful dog named Hachiko, who dutifully walked with his owner to Tokyo's Shibuya train station every day, waiting patiently there for the master to return from work. When his owner died suddenly one day without returning home, Hachiko continued to wait for him, returning every day for ten years before passing away in the same spot where he had last seen his master. I've traveled to Tokyo dozens of times and frequently stay at a hotel across from what is now universally known as the "Hachiko entrance" of Shibuya station. Even from twenty-five stories up, I can look down and see the bronze life-size statue built on that spot in memory of the remarkable canine. Hachiko's story has achieved such mythic status that I'd wager nearly every adult in Japan—and most Japanese schoolkids—knows at least the basic outline of the story and its message of honor, duty, and faithfulness. And if you say to literally any native Tokyoite, "I'll meet you

The story of Tokyo's faithful dog Hachiko reinforces the Japanese virtue of steadfast loyalty.

in front of Hachiko," they'll know exactly what you mean. Hachiko died seventy years ago, but his story still has a long time to run.

The lore of a company is a potent way to communicate values and objectives across a widely dispersed and multicultural organization. Hewlett-Packard's tale of starting in the garage is not only cherished by HP's hundred thousand staff members around the world, but is also inspiration for entrepreneurs everywhere who are starting on a shoe-string but aspire to future greatness. And Michael Dell's variation of launching toward multibillionaire status from his college dormitory provides extra reassurance for entrepreneurial dreamers who don't have ready access to the requisite garage. Southwest Airlines people love to tell you how the idea for the company was scribbled out on a cocktail napkin, and thousands of other budding business plans have been started the same way since.

The Right Story at the Right Time

Storytelling expert Stephen Denning reminds us that not just any story will do. In books like *The Leader's Guide to Storytelling*, he helps us match up the right kind of narrative with the right situation. Denning says that business stories have focused purposes like sparking action, transmitting values, fostering collaboration, or leading people into the future. Before you begin a story, it's important to know what specific out-come you are hoping to attain. For example, when my kids were toddlers, I used to read them a bedtime story most nights with the intent of calm-ing them down and helping them go to sleep, but if you get that outcome in a business setting very often, it's time to work on your Storyteller role.

Regardless of which type of story you're telling, Denning urges busi-ness leaders to be conscious of the distinction between "true" and "authentic." He says corporations spend too much time speaking at the boundaries of truth, when they should aspire to stay at the heart of authenticity. Truth-in-advertising laws insist that the majority of mes-sages you hear from companies are true—at least in the narrow sense of the word—but authentic is more like "the whole truth and nothing but the truth." He distinguishes between the two with a quick historical illustration: A true story would be *"Titanic* sails on maiden voyage. Seven hundred 'happy' passengers arrive in New York." Technically true, yes, but it fails to hold up to any further scrutiny. True stories can have a lot of spin to them. Authentic stories have deep integrity. Customers, employees, and members of the world community can tell the difference.

> Denning urges business leaders to be conscious of the distinction between "true" and "authentic." He says corporations spend too much time speaking at the boundaries of truth, when they should aspire to stay at the heart of authenticity.

Recently, I've noticed a new trend in company mythology. Whereas most older company lore focused on the visionary founder or president, increasingly stories are being told about the little heroes that make up a company's day-to-day operations. One afternoon at the Starbucks across from our San Francisco office, I was intrigued to find a compelling page-long story neatly laminated and prominently displayed at the front counter. There was a photo of a San Francisco Starbucks barista who had become a manager, and a story in her own words. It told of hard work, joy at supporting good causes like the fight against breast cancer, and her ability to find rewarding challenges in her job. The woman noted with pride that her Starbucks salary and stock had enabled her to buy a home in the San Francisco Bay Area, no small feat. I don't know if you're a Starbucks fan or not, but this small story struck me as authentic and compelling. Of course, it was a wonderful recruiting tool, but it also implicitly says to coffee drinkers everywhere, "We're good people—the kind of people you would be if you were running the world's biggest collection of coffee stores."

No matter how small or large your company, the organization is constantly collecting and spreading stories about your business, your values, and your achievements. Mythic stories endure because they become shared symbols. Passed along from person to person and generation to generation, myths do not always preserve all their factual detail, but the best myths have a ring of authenticity and tell an underlying truth.

Think about the myths you tell, and always strive for authenticity.

Tell Me a Story

So where does storytelling begin? One answer to that question is to ask where Storytellers find their inspiration. And why they believe storytelling is critical to innovation.

As readers of *The Art of Innovation* may remember, Jane Fulton Suri is IDEO's thought leader in the human factors discipline that inspires the Anthropologist. So what can an Anthropologist teach us about the role of Storyteller? Plenty. Jane fervently believes that her work is largely based on listening to and interpreting human stories. On writing them down. On seeking the underlying meaning and implications. On hearing both text and subtexts. On being a Storyteller.

Many of us make the mistake of trying to take shortcuts on the way to capturing other people's stories. With a "bottom line" mind-set, we say, just give me your insights. Tell me the highlights. We're looking for bullet points in a winning presentation. We're looking to cut to the chase.

Jane doesn't ask for instant insights and she doesn't jump to conclusions. She doesn't ask yes-or-no questions, either. She goes into the field and finds interesting people (and almost everyone seems interesting to Jane). Instead of asking questions like "What do you like or dislike about your mobile service?" Jane will start with "Tell me a story about a time your mobile let you down." In the ensuing conversation, she'll uncover plenty of likes and dislikes, but she builds a better personal connection and gains deeper insights by basing the discussion around stories. To her mind, it's about respect and humanity. Asking for a story celebrates and authenticates the experience. She knows from years of fieldwork that when she does this right, without pretense, her subject is thinking, "Oh, they want to listen to me!" Everyone wants to be listened to, and if you can tap into a reservoir of personal stories, the insights you're seeking will start to emerge.

Why is having this patience and laying this groundwork of trust so pivotal? Because, as Jane puts it, storytelling just happens to be a fundamentally human way of conveying information. There's a reason why folklore and religious stories endure. Storytelling is part of the fabric of humanity. And when you respect storytelling, you acknowledge that you're engaged in a human enterprise. You elevate your work. You create a common language. You begin to build a larger community.

Ask Jane about the importance of storytelling and, not surprisingly, she'll tell you a story. A few years ago, as Jane recounts, we worked on a project for a hospital. We kicked things off with a meeting of about twenty nurses, administrators, and doctors. That first session of a Transformation program is typically filled with a healthy mix of skepticism and anticipation among the assembled participants. Start out on the wrong foot and the whole project can suffer. But this particular day, the power of storytelling helped the team bond and reinforced the significance of our goal. In preparation for that kickoff meeting, Jane had asked each member of the group to do a little personal homework: recalling a really bad or good health care experience they had witnessed firsthand. Something *personal*.

Within minutes of going around the room, people were laughing. And crying. One nurse recounted an intense day when a dying man asked her to call his wife, but in spite of all her efforts, the nurse couldn't locate her. She was frantic. The patient was slipping away. She had to find his wife. The man grabbed her arm. "It's OK," he told her. "Now we've got something to do together. You're going to teach me about dying," he said, "and I'm going to teach you about living." The wife never arrived, and the nurse realized that some part of her role that day wasn't just about trying to save this man. Instead, she could offer a priceless gift, letting him share his last moments on earth with another human being.

> "It's OK," he told her. "Now we've got something to do together. You're going to teach me about dying," he said, "and I'm going to teach you about living."

As the nurse finished her story, there wasn't a dry eye in the room. Many of these people hadn't known one another before that morning. But the stories brought them together and infused the project with passion and insight. The collection of stories emphasized how wide the range is between very bad experiences and very good ones, and reminded them how high the stakes are in getting health care services right. There's nothing like stories to connect you with a subject, to pull a team together to work on human issues in a human way.

IDEO has worked alongside Minneapolis-based Medtronic for many years—a blue-chip medical technology company that's best known for its market-leading cardiac pacemakers. Medtronic employees are paid well, and many of them are shareholders, so that should be enough to motivate them to do a good job, right? Well, yes, but Medtronic is looking for more than a good job; they want to out-innovate the competition. And they use storytelling as a tool. One of their senior executives told me that whenever the team needed a spark, they would simply bring in patients—or the children of elderly patients—and say, "Please tell us a story about how a Medtronic product changed your life." The result, my friend at Medtronic told me, is positively electric. Because after a few of these life-renewing stories, even the "tough guys" in the room start to get a little teary-eyed, and the entire Medtronic team goes back to work afterward, inspired and motivated to do their absolute best for people like the ones they just met.

No, we aren't always working to save lives or comfort the dying, but most of us believe in the value of what we do. Go out and find some real people. Listen to their stories. Don't ask for the main point. Let the story run its course. Like flowing water, it will find its own way, at its own pace. And if you've got patience, you'll learn more than you might imagine.

Dreaming Up a New Story

Introducing change in a large organization can be tough. It's not enough to dream up new concepts. Sometimes you've got to dream up a new way to tell the story.

A few years ago, we had a challenging assignment from a major car manufacturer. We began with the premise that the automotive industry sometimes considers women as an afterthought. Surprisingly few cars are designed or sold with women in mind, despite the fact that most purchasing decisions are made or influenced by women. With that clear opportunity, the auto company asked IDEO to plumb new-car ideas for women in their twenties.

We dove into the project with our usual enthusiasm. We recruited female staffers and friends to shop till they dropped at retailers like Urban Outfitters and Origins, immersing themselves in the culture of women's shopping. Of course, they didn't really have to buy. We gave a number of women a fictional budget and sent them out to dealers to "buy" a new car. The dealers ranged from ambivalent to downright nasty. They intimidated the women and tried to take advantage of them. "Come back with your daddy," suggested one sexist auto salesman. Interviewing the women, we learned a few things. The most obvious was that while many women love to shop, most car dealers make them feel miserable. A major lost opportunity. Another obvious discovery was that women in their twenties apparently adore convertibles—eleven of our twelve women had either convertibles or sunroofs in their cars.

But pushing farther below the surface, we found something more subtle. The shopping experiences at home decor shops demonstrated that the women were drawn to something more fundamental than convertibles. What they sought was a certain lightness and whimsy—qualities not found in most modern cars. Meanwhile, our project space for this exploration filled with images and props, everything from fun T-shirts to magazine collages—shots of active young women and heart-throbs, as well as the latest shoes and fashions. Our team got excited about emerging concepts and was anxious to share our findings with the client. But somehow, delivering the standard wire-bound report in a plastic cover seemed flat, given the lively material we were trying to convey.

"Wouldn't it be fun if our report were like one of these magazines?" suggested one of our team members, flipping through one of the dozens of women's magazines that had accumulated in our project space. The attraction was obvious: The chatty, familiar, and casual editorial. The bouncy self-help tone. The sexy design, emphasizing photographs and design over text.

We'd never done it before, but the team got excited about breaking new ground. They modeled their effort after several young women's magazines, especially *Lucky*, with its perfect tag line, "It's all about shopping." Our "magazine" told the story of what we'd learned about women and cars from our interviewees, observations, and brainstorms. There were personal features with snapshots of our dozen interviewees: who they were, what car experiences they'd had, how they viewed the automotive world. An article on transitions explored one of our major findings—that these women were all going through major lifestyle shifts, from single to married, from adolescent to adult, from club-hopper to mom. An article on influencers described the trusted sources these women turned to for advice or help with life's challenges. Of course, we had great photos of women going through the buying process (our resourceful recruits had no problem inventing a story for why they wanted to visually chronicle the process).

What did our carmaker think? We literally couldn't print enough copies of the thirty-two-page glossy magazine. The company found it different and fun and thought-provoking. It was warmer and more personal than what they were accustomed to. Most important, it raised questions about the critical attributes women found appealing. But our storytelling wasn't complete. As we honed our findings, we settled on key attributes. Concepts like "Haven," the idea that for a young woman just out of college, still living with roommates, the car may be the only place to escape or find quiet comfort. As the project progressed and the concepts became more refined, we searched for a new medium for presenting our ideas. We settled on the metaphor of a user's manual, much like the manual in a car's glove box. A tool kit for designers that explored our vision of the attributes and features women would like to see in a car.

We capped off the project with a four-minute animated video, the story of a woman's car-buying journey—researching online, visiting the showroom and picking out features, and finally, blasting off on a road trip with a roommate. A self-help magazine, a car manual, an animated video—three very unique and focused storytelling vehicles that helped craft our observations and findings into a more actionable form. Don't forget the importance of telling a story. Every team can benefit by being open to considering just what medium might best convey your story. You may not believe the old Marshall McLuhan adage that "the medium is the message," but the right medium can certainly support and amplify your intended message. Just as you craft your message, give attention to what medium is most likely to get your point across.

Don't Zap That Infomercial

Sometimes stories work best if they shock. So here's what might sound like a shocking idea: If you're looking for new clues to better corporate storytelling, consider a radical source—the much-maligned infomercial.

I haven't lost my mind. I know infomercials have a dicey reputation. But a creative Storyteller keeps an open mind and is willing to find learning in unusual places. Like them or not, television infomercials are a very focused form of modern storytelling. They're phenomenally successful, and they are *not going away*.

The best infomercials work because they build a detailed, persuasive case for the product. It's the same thing companies must do all the time, whether they are advancing a corporate agenda or introducing new services and products to business partners. So what makes a successful infomercial? Johann Verheem, founder of the infomercial firm EQmedia Partners, says you have to build to three mini climaxes before the grand finale. You introduce skepticism or controversy, air common doubts and worries, then knock down the objections one by one. Snicker all you want, but most corporate videos aren't nearly as comprehensive or persuasive as infomercials. Three minutes into a corporate video, they're often still showing how many square feet of warehouses they own, while the infomercial is telling a story that's building dramatic tension.

Part of what has allowed infomercials to "tune" their response rate so successfully is that they have an enviably quick feedback loop—the instantaneous spike of calls into the 800 numbers where "operators are standing by." And no infomercial illustrates the power of instant feedback better than the story of the George Foreman grill. When first introduced by Salton without its celebrity endorsement, the new barbecue grill was met by lackluster sales. Even when retired boxer George Foreman made the first commercial, it was still not an instant success. Then, in a candid, informal moment, Foreman grabbed a burger off the grill and took a hungry bite, seemingly not conscious of the camera. When that authentic moment was introduced into the next airing of the infomercial, the switchboard lit up like Times Square. Sales went through the roof, and that *Candid Camera* moment has appeared in every infomercial for the product since.

Of course, I'm not suggesting that you directly mimic infomercials. But the reality is that a lot of organizational storytelling remains stuck in the corporate-video syndrome—flat, doubt-provoking, and occasionally mind-numbing renditions of company objectives. They desperately need help. So if you've got a high-stakes story to tell and a limited time to do it, screen some great commercials, films, and yes, even infomercials. Take what works, lose the rest, and come up with something distinctive that brings new energy to your stories.

Beyond the Fortune Cookie

Great stories work in all shapes and sizes. The power of storytelling is what drives the sales of millions of fortune cookies every year. Otherwise, it would be hard to understand the runaway success of fortune cookies. Let's face it: They're not on any food critic's list of "top ten taste treats," and they have a texture like injection-molded plastic. There's nothing wrong with their classic look, but it certainly doesn't promise culinary delight or even comfort-food indulgence. So what explains their enduring popularity? Why, the *fortune*, of course. Fortune cookies are about 10 percent cookie, 90 percent experience, and we all love the ritual of figuring out who belongs to which cookie, then breaking them open with a crisp, satisfying snap and reading the fortunes aloud to everyone at the table. It's simple. It's fun. It's a shared experience.

So have you noticed the fortune-cookie phenomenon spreading to other venues? The first place I spotted fortunes without the cookie was at Palomino restaurants (there's one across the street from IDEO San Francisco). Just like in a Chinese restaurant, Palomino presents each diner with a fortune when they deliver the bill (presumably as a way to soften the blow). The fortune cards at Palomino required a little work to open—tearing off three perforated edges—which was part of the fortune-cookie-like ritual. Their fun little fortunes give closure to the meal as, one by one, diners share the bits of wisdom inside.

Having jumped from East to West, fortune-cookie wisdom then jumped from food to drink. In case you haven't noticed, most high-end juice and iced tea drinks—especially the ones in glass bottles—use the inside of their metal caps as a tiny media opportunity. Honest Tea taps into wisdom of the ages like a fortune cookie, though the quotes are a bit more contemporary than Confucius. Last time I drank an Honest Tea, the sage they quoted was Lily Tomlin, who reminds us, "Even if you win the rat race, you're still a rat." The few words in these fortunes would barely qualify as a story, but they are often conversation starters, setting off a chain reaction of personal stories among friends and colleagues.

Jumping on the storytelling-anywhere bandwagon, IDEO suggested it might be possible to print riddles or interesting facts on a Pringles potato chip using food-based inks. Sure enough, P&G came up with a clever way to do it, teaming up with the folks from Trivial Pursuit for some of the content. It's a small step in the chain of innovations that sprang from the fortune cookie, but it did turn eating Pringles into a fun social event. And it increased their market share by 14 percent in the first year.

Can you use storytelling or teaching on a micro level to cement the bond with your customers? Could your vehicle entertain or teach me a little something every time I start it up? Could your elevator tell me a story that I could use in today's meeting? Could your mobile network connect me to just the right version of a dial-a-story? Because even the smallest stories can go a long way toward making your service or product a little more exceptional.

Seven Reasons to Tell Stories

Why should organizations care about becoming better Storytellers? Well, we've given that question a lot of thought. Roshi Givechi of our

San Francisco office even formed an IDEO advocacy group to share our latest insights on how storytelling aids innovation projects.

Here are seven reasons she believes organizations should care about becoming better Storytellers.

1 **Storytelling builds credibility.**
   We often tell stories about firsthand field research during our initial client meeting. Even though the client may have decades of experience in their industry, the immediacy of our first-person narrative (drawn from recent observations) offers credibility, even if—maybe *especially* if—it goes against the client's sense of the market. Passion and a fresh perspective demand respect. The client may know a particular market, but the Storyteller with an intriguing first-person narrative is the world expert on their own experiences.

2 **Storytelling unleashes powerful emotions and helps teams bond.**
   Captivating stories trigger emotional responses that frequently spark valuable insights. As I mentioned earlier, we'll often kick off a project by asking clients to tell stories about particularly positive or negative service experiences. We've had participants cry when talking about extraordinarily good—and noticeably bad—encounters. By the time everyone's had a laugh or a nod of recognition, the team is stronger and more focused. You'd be surprised. Even executives fond of a left-brain, analytical approach can get pretty fired up about a story that cuts to the essential human questions.

3 **Stories give "permission" to explore controversial or uncomfortable topics.**
   Sometimes we invite team members to try a little show-and-tell: to bring in a favorite object and tell us a story about it. One client recently brought in a mountaineer's ice pick to symbolize his belief in his product's need for dependability, reliability, and trust. Most likely, he would have been uncomfortable discussing such "touchy-feely" concepts in the abstract, but in the

context of a story, it felt very natural to do so. Storytelling can act as a kind of Trojan horse, getting past our initially defensive reactions of doubt or skepticism, enabling us to have an open discussion about a relevant idea.

4 **Storytelling sways a group's point of view.**
We aren't opposed to studying demographics, industry dynamics, and market trends. But facts alone provide little direction or inspiration for a new project. A compelling story can serve as a parable that helps shape a group's perspective. Most great leaders have used storytelling as a part of their strategy for success—in ancient times around evening fires, and now using all the modern communication options that are available to them.

5 **Storytelling creates heroes.**
The observations that inspire so much of our innovation work are often grounded in the stories of real people—customers or would-be customers with needs that aren't met by today's products or services. These individuals give a name and often a face to the design objectives of a project. You'll often hear team members say, "Would that help Lisa?" Sometimes we'll combine elements of these real people and, as in the movies, create a composite embodying most of our objectives in one fictionalized character. These characters give us a hero—someone to innovate for.

6 **Storytelling gives you a vocabulary of change.**
One of my goals when I speak before corporate audiences is to introduce new concepts and spark new conversations that encourage innovation. Many of the best business books of the past twenty-five years have introduced new phraseology into the boardrooms and meeting rooms of the world. For example, Malcolm Gladwell popularized the phrase "tipping point" in the late 1990s, Clayton Christensen gave us "disruptive technologies," and Geoffrey Moore got businesspeople talking about "crossing the chasm." At IDEO, our best stories are generously

seeded with phrases and words that provide new frameworks for innovation efforts. Internally, we talk about "T-shaped people" using "design thinking" to come up with breakthroughs in "Phase zero" of a "think-to-build" project. Some of a firm's language is self-explanatory, while other expressions are a bit obscure, but the words in our stories reinforce concepts and accelerate the diffusion of innovation. Language is crystalliza- **Good storytelling cuts** tion of thought, so the stories **through the clutter.** matter, and so do the words.

7 **Good stories help make order out of chaos.**
We have too many items on our to-do list, too many voice-mail messages, and too many unread e-mails. We cope in part by developing a protective form of attention-deficit disorder that allows us to jump from subject to subject—to screen out, ignore, neglect, or actively forget what would otherwise overwhelm. Good storytelling cuts through the clutter. Think back a few years: Most likely you will have trouble remembering a specific e-mail or phone conversation. But I bet you can remember a good story told to you by your parents long ago, or by your first boss or your best friend. Telling stories is one of the ways you begin building a relationship—whether it's in life or business.

Touring Tales

Finally, don't forget that one of the best opportunities to tell a story is all around you. Your physical space has the potential to communicate important stories in ways that no PowerPoint presentation could ever match.

An IDEO tour for clients or invited guests is a common occurrence. It's not quite like the Universal Studios tour, but apparently it does have some information and entertainment value. I've led more than a thousand client tours myself in the past eighteen years, and the feedback is that tours are worthwhile. What makes up a tour? IDEOers take a handful of guests around the campus, using the spaces and artifacts they encounter to trigger relevant stories, in the way that "story beads" help guide Native American storytellers through a series of tales. We've got several "storytelling entrances," places you'll find some of the favorite products and services we've worked on during the years, as well as our Materials Bar and Tech Box. There are waypoints along the way that represent case studies, pieces of the creative process, or insights we think are worth sharing. During the tour, we also wander through the eclectic offices, giving visitors a sense of how we work and what we do.

Why does the tour remain so rich and vibrant? I used to think it was our ever-changing space, which admittedly doesn't resemble most company offices, and the wide range of colorful and visual examples of our work. More recently, I realized that it stays fresh mainly because there's no script, no set path, no official tour stops. It's entirely an oral tradition. Of my thousand or more tours, no two were quite the same, and my colleagues give a tour that is different again from any of mine. We believe you shouldn't let one or two specialists deliver all your tours—whether your company makes teddy bears or computer chips. You need a wide range of personalities if you want your storytelling to remain varied and fresh.

A good tour not only shows off your company's accomplishments but also can enrich your culture of storytelling and boost your team's morale. There's something special about working in a place distinctive enough to make outsiders want to come and have a look. And that's part of why the storytelling value of a good tour can't be matched by mere PowerPoint. There's no substitute for a firsthand experience.

During the last several years, I've noticed that companies are getting markedly better at setting the stage for their tours. A while back, I got a guided tour of Sony's Media World atop their headquarters building in Shinagawa, which I would rate among the best office tours in the world. It's like a year-round trade show of the near-term future in consumer and "prosumer" electronics. I admit, however, that I was a bit disappointed when I went back two years later and neither the technologies nor the stories seemed to have changed. The problem with investing so incredibly much in the tour is that they have to amortize their invest-

Tour stops can provoke great storytelling. Sprinkle your space with show-and-tell opportunities, and you may discover your team is full of Storytellers.

ment over a long period of time. Our tour is much less polished, but our minimal investment allows us to tinker with it continuously.

Once you've built a stage, you need talented and enthusiastic people to deliver your tour. If it's fun, if the stops on the tour and the staffers you encounter give you a charge, you won't have any trouble recruiting and retaining tour guides. My biggest suggestion is to allow each Storyteller the freedom to shape their own tour, with their own favorite stops and little asides. That way the thousandth tour may be as spontaneous and inspired as the first. And you'll find the tour takes on a life of its own.

No matter what stories you tell—whether they are tours of your cool offices, or video prototypes of new services you're developing, remember the first rule of Storytellers: Keep it authentic and entertaining. Strike an emotional chord. Make it a story people will want to pass on. Because stories are a part of your personal legacy and the essence of your authentic brand.

CHAPTER 11
In the Mix

'm a firm believer in the galvanizing power of personas. Adopting even one new role can bring both cultural and business benefits to your organization. But the real payoff comes when you gather several roles together and blend them into a multidisciplinary team. Innovation is ultimately a team sport. Get all the roles performing at the top of their game and you'll generate a positive force for innovation.

Successful teams don't need to be best of class at the skills and tools of every persona. As in an Olympic decathlon, the objective is to achieve true excellence in a few areas and strength in many. If your team excels at the fieldwork of the Anthropologist, maybe you can get by without having a world-class Set Designer. Similarly, the insights you gain from an Experimenter's practice of trial and error may offset the need for a strong Cross-Pollinator. If there are ten approaches to creating and sustaining an innovative culture, what counts is your total score, your ability to regularly outperform the competition in the full range of daily tests that every company faces.

Winning at Innovation

Perspiration and old-fashioned hard work have long been recognized as central to the process of creativity. Even gifted athletes remind us that it takes disciplined effort to make the best of their talents. Teaming up for innovation is a lot like an athletic event, and many of the same principles apply:

1   **Stretch for strength.** In the long run, flexibility is more important for your organization than size or even power. Although companies like Bethlehem Steel, Pan Am, and Montgomery Ward were giants in their heyday, they lost huge chunks of market share to companies that overcame power with responsive new business models. Personas like the Cross-Pollinator and the Experimenter help keep your organization nimble and fresh. Flexibility is the new strength.

> Flexibility is the new strength.

2   **Go for distance.** Innovation is not just a program. It's a way of life—just as the ultimate wellness program is not a fad diet but a healthy lifestyle. Adopting the personas and nurturing a culture of innovation is not a job to be delegated to the marketing department, or HR, or R&D. The spirit of innovation permeates the entire enterprise at great companies. Keep your innovation personas in condition and encourage your team to do the same.

3   **Never surrender.** Top athletes have a tireless never-say-die approach to their daily workouts and their competitive performance. The Hurdler recognizes he will have to overcome a series of obstacles, even if only one or two are visible at the onset. The Director knows that a steady pace of new ideas is better than a single burst of energy around a new initiative that gets launched with enthusiasm but abandoned a few months later. The Collaborator can multiply your energy reserves through cooperation with other internal and external teams, extending your organizational stamina by sharing the load. Most of the innovation personas boast the same relentless attitude, keeping their point of view in motion through a process of continuous learning and persistent advocacy.

4   Embrace the mental game. Almost any successful athlete will
    tell you that the mental side of their competitive spirit looms
    large, especially when fatigue and frustration set in. What neg-
    ative behavior patterns or views hold you back? What blind
    spots can you eliminate? It takes supreme mental focus to make
    that final pole vault, after missing two attempts, while the whole
    world is watching. In the same way, all the innovation personas,
    especially the Hurdler and the Experimenter, need mental tough-
    ness to keep going when common sense and physical fatigue may
    say it is time to stop. Innovators have the *uncommon* sense to
    pursue promising ideas long after their colleagues would have
    given up or given in.

5   Celebrate coaches. Even in solo performances, most high achiev-
    ers have a great coach who believes in them. And while the right
    coach can catapult you to greatness, the wrong coach—or some-
    times no coach at all—can limit the trajectory of your career and
    your life. If your coach is like the legendary high jumper Dick
    Fosbury's early trainers, consistently trying to discourage your
    innovations, maybe it's time to find a new mentor. Seek out some-
    one you trust who will nurture the Anthropologist or the Col-
    laborator or the Storyteller in you. The right coach will bring
    out the best in you, and you'll notice the difference.

When putting together a team—in sports or in business—you don't
want to rely too much on one star player. Seek a rich mix of personas and
personalities. Sure, the differences among their talents and points of view
will cause some friction every once in a while, but a little creative abra-
sion can be productive when you are pushing for continuous innovation.
So take a good look at your team's composition. Where is it strong and
where is it not quite ready for prime time? Do you need to nurture or
recruit a new role? For example, are there untapped learning opportunities where the Anthropologist or the Cross-Pollinator could help? Where
could a good Experimenter enable you to screen or validate more ideas
through enlightened trial and error? Could you use a Hurdler or a Collaborator or a Director to keep new initiatives moving forward? Could a

Set Designer bring out new energy in your team or create great customer experiences in your locations? Are you fully leveraging your internal and external Storytellers? Luckily, you don't have to find all the innovation personas in one individual. You can field a whole team of people that collectively win for your organization. Bring out the best in your innovation team and you can all share in the success.

Putting the Personas in Context

It is my fervent hope that this book will spark lots of productive conversations that lead to action. I hope that the ripple effect of those conversations will encourage companies to build or renew a culture of innovation that promotes organic growth. When applying the personas to successful, ongoing organizations, it's important to note that the innovation personas do not necessarily replace existing roles and titles. In the last few years, I've noticed that many companies have Innovation Directors, and I've even run across some business cards with the title Experience Architect, but that's not critical to the effective use of personas. We've got plenty of Experience Architects and Cross-Pollinators at IDEO, for example, but you never find those roles printed on their business cards or written into job descriptions. The personas can comfortably live side by side with traditional titles. Labels like engineer, programmer, project manager, and executive are not likely to go away anytime soon. You may be an engineer with the heart of a Cross-Pollinator and the spirit of an Experimenter. At your best, there's a synergy that comes from bringing these talents together.

You may not realize it, but I'd bet you've already piled roles on top of your existing roles. When I became a parent, that suddenly became my most important role. It certainly consumes a lot of time. But I didn't give up any other roles, like husband, brother, or IDEOer. None of these roles negates the others, and like most people, I am constantly juggling roles. When things are going well, the roles can be wonderfully complementary.

The message is that it is possible—even desirable—to blend a traditional, discipline-based role with an innovation persona. You can be an Anthropologist even if your business card calls you a systems analyst. You can be a Cross-Pollinator in the Marketing Department. You can be a Hurdler in Accounts Payable. A Set Designer in Human Resources. A Storyteller even if your degree is in finance. Don't let a title or job description hold you back. Show me a list of people who changed the world, and I'll show you a group of people unconstrained by traditional roles.

Remember the idea of well-rounded, T-shaped people that is so intrinsic to IDEO's strategy for hiring and professional development? Maybe for you, your personas can be the bar of your T. If you have spent years building depth of experience in your functional area, then becoming a Collaborator or a Caregiver or an Experimenter might give you breadth. We believe the future belongs to T-shaped people. And it's not easy to replace a T-shaped person. The broader your talents, the more your ability lies in the overlaps between disciplines, the less likely you will find yourself outsourced.

While I believe in all ten of the personas, you don't need every persona on every project, and certainly not at every moment. A good analogy might be the tools in a carpenter's toolbox. You seldom need all the tools at once, but the perfect kit of tools is a set where you use all of them pretty frequently.

Innovation doesn't happen on its own, but with the right team, you're up to the challenge. So find new paths of learning with the Anthropologist, the Experimenter, and the Cross-Pollinator. Organize for innovation with the Hurdler, the Collaborator, and the Director. Ask the Set Designer to help build your stage, and bring on the Experience Architect, the Caregiver, and the Storyteller to wow your audience. Innovation doesn't just turn companies around. It becomes a way of life. It's fun. It's invigorating. It works. With all ten personas on your side, you can drive creativity through the whole organization and build your own unique culture of innovation.

Good luck!

INDEX

Introduction
p. 1   ©Hunter Lewis Wimmer/IDEO

1: The Anthropologist
p. 15   Courtesy of Mauro Campagnoli
p. 20   Method Cards ©Mark Serr
p. 24   Margaret Mead ©Corbis
p. 27   Grandmother and child cooking ©Michelle Lee/IDEO
p. 32   Universal News Courtesy of Universal News & Cafe
p. 35   Turnstile ©Steve Murez and IDEO

2: The Experimenter
p. 41   ©Andy Sotiriou/Getty Images
p. 45   Balance Beam ©Nicolas Zurcher
p. 46   Gyrus ENT Diego prototype ©IDEO; product ©Rupert Yen
p. 54   Paper prototype ©Nicolas Zurcher
p. 58   bmwfilms.com, Courtesy of Fallon Worldwide, Clive Owen and cityscape
           photography by Michael Crouser, BMW photography by Mark LaFavor
p. 63   Collaborating Kids ©Kara Krumpe/IDEO

3: The Cross-Pollinator
p. 67   ©Morgan Mazzoni/Getty Images
p. 69   Frisbee and Pie tin ©Nicolas Zurcher, thanks to Richard Pancoast
p. 76   T-shaped people ©Hunter Lewis Wimmer/IDEO
p. 81   Fosbury Flop ©Corbis
p. 85   Muji products ©Nicolas Zurcher
p. 88   Nihls Bohlin, Courtesy of Volvo Car Corporation

4: The Hurdler
p. 91   ©Hoby Finn/Getty Images
p. 98   Virgin seat back video system©Virgin Atlantic Airways
p. 102   Edwin Moses ©Corbis
p. 105   Graphing Calculator, Courtesy of Ron Avitzur
p. 107   Zinio ©Hunter Lewis Wimmer/IDEO

5: The Collaborator
p. 113   ©Phillip Lee Harvey/Getty Images
p. 123   Packaging Hall of Fame ©Sean Ferry/IDEO
p. 126   Samsung SimpleMedia, ©Rick English Pictures
p. 128   Collaborating Triathlon, Courtesy of Cargill, Inc.
p. 130   Baton Pass ©Lori Adamski Peck/Getty Images
p. 136   Soccer Pass ©Corbis

And special thanks to Mauro Campagnoli for sharing the photo on page 15 from his anthropological work with the Baka Pygmy culture.